# Introduction to the Biochemistry and Physiology of Plant Growth Hormones

Introduction to

# The Biochemistry and Physiology of Plant Growth Hormones

## I. D. J. Phillips
*Department of Biological Sciences,*
*University of Exeter, Devonshire, England.*

## McGraw-Hill Book Company

**New York** · London · Toronto · Sydney

# Preface

A fertilized egg of a flowering plant is the starting point of a complex series of events which lead to the appearance of a new mature plant. These events are collectively referred to as plant development.

Development in higher plants is a regulated process, and to a large extent the pattern of development is 'pre-ordained' in the sense that inherited genetic information serves as a series of instructions which direct the course of development. This means that DNA-encoded information is expressed during development, and that different parts of the message are utilized at different times during the ontogeny of individual cells, tissues, and the whole organism.

However, gene expression is modified by influences which arise outside the cell. In other words, a developing cell responds to its environment within the plant body and, in turn, to the external environment, since the latter may be expected to affect the internal environment of the plant. It has become increasingly clear over the past ninety years that the rate of growth, and pattern of differentiation, in a plant cell are susceptible to control by chemical agents which arrive in the immediate environment of the cell after having been translocated from other parts of the plant. These chemical agents have been given various names, but are commonly termed plant growth hormones. They are extremely important in the integration of developmental activities in spatially separated regions of multicellular higher plants. Plant growth hormones are also concerned, as we shall see, in the responses of plants to the external environment.

Because plant growth hormones are essential to the control of plant development, they have been the subject of intensive research for many years. Courses intended to teach something of plant morphogenesis, or developmental physiology and biochemistry, have, of necessity, to take into account the known plant growth hormones. Similarly, students of agriculture, forestry, and horticulture also need to understand the biochemistry and physiology of plant growth hormones, partly because of the very great practical importance of a number of these substances or their analogues.

As its title suggests, this book is intended to provide an introduction to the nature and functions of plant growth hormones. It is believed that the arrangement of the subject matter, and level of treatment are such as to give even a newcomer a wide-ranging and contemporary view of this fast-developing field of research. The vigorous condition of research into all aspects of plant growth hormones and their activities is, in fact, a very real justification for this book, for only a relatively small and inexpensive book can be regularly updated. This means that future generations of students, no matter what their speciality in basic or applied plant science, will be able to obtain a contemporary introduction to the subject of plant growth hormones.

Although this is intended to be an introductory account, it is hoped that it does not contain unwarranted generalizations. As far as possible, the methods employed in elucidating problems are stated, and their limitations pointed out. Only very rarely can one give an unqualified explanation of the effects of plant growth hormones in a particular developmental phenomenon, but this should in no way detract from the interest in the problem. In fact, the inability to provide pre-digested conclusions should act as a stimulus to the student. Contradictory results and/or 'conclusions' *are* produced by different workers, but this situation is accepted as normal by scientists. The exchange of criticism between researchers is an indispensable part of the pursuit of elusive 'facts'.

Some readers may regret a lack of literature citations in the text. However, since this is neither a history book, nor a reference work for advanced students, then it is more important to focus attention upon the message rather than the medium. Although in some circles it is fashionable to claim that the medium *is* the message, such a view has no place in science. What is important is that students should acquire a contemporary outlook on the subject, and become aware of the directions in which the most important advances are occurring. Nevertheless, each chapter is followed by a list of references, which have been carefully selected to provide the interested reader with strategic entries into the very extensive literature which exists on the subject of plant growth hormones. In addition, as most of the figures have been prepared from original research papers, and the source of each is given, then these also provide a means of identifying cited work.

Finally, I wish to acknowledge all those research workers whose efforts have provided the foundations for this book. I particularly thank those who have been kind enough to help directly, by the provision of photographs and data which are reproduced here.

*I. D. J. Phillips*

# Contents

# Contents

# 1. The nature of plant growth hormones

'My, how you've grown!' Most of us will have endured, in our earlier years at least, that fond but curiously irritating greeting from far-flung relations. Irritating, perhaps, because to state the obvious usually appears absurd. 'Of course I've grown', we may have thought (but only the less circumspect actually dared to say), knowing full well that the passage of time in early life is always accompanied by an increase in body size. Similarly, we all know that plants grow. Tempting though it may be so to do, we nevertheless cannot define *growth* as getting bigger. Even in its broadest sense, we must qualify such an over-simple definition—every girl is well aware that getting bigger in the wrong places doesn't necessarily constitute growth! She knows that, with luck or will power, she may reverse the unhappy trend. Thus we may, simply, define growth as an *irreversible* increase in size, mass, or volume. This crude definition relies solely upon the morphological expression of growth activities, but, as with girls, there is more to growth than that which immediately meets the eye.

In autotrophic higher plants, growth basically consists of the conversion of relatively simple inorganic substances (water, carbon dioxide, and mineral elements) into increasing quantities of proteins, carbohydrates and, to a lesser extent, fats. The consequent increasing bulk of the plant body may be seen and measured from outside. Internally, growth involves more than just the accretion of ever-increasing quantities of proteins and carbohydrates, for cellular changes also take place. Multiplication of cells by mitosis, and enlargement of individual cells during vacuolation both constitute

1

fundamental components of growth. *Growth*, therefore, is a term commonly applied to all *quantitative* changes that occur during the life of a plant.

However, plants not only grow during their life, but they also *develop*. Plant development is characterized by *changes in form* in the plant body, and is brought about by successive patterns of *differentiation*. This latter term is applied to all the *qualitative* differences which appear between cells, tissues and organs during growth. Such differences may be structural, for example, phloem sieve tubes, companion cells, xylem vessels and tracheids, and epidermal cells, all arise as a result of differentiation of these cell types from originally identical cells formed at the shoot apex. In other examples, differentiation may not result in the appearance of morphologically distinct cell types, but nevertheless morphologically similar cells may possess different biochemical properties (e.g., some parenchymatous cells actively convert sugars to starch, whereas others do not store carbohydrates in this way).

Separate categorization of growth and differentiation processes is a purely artificial device. Both go on simultaneously in the same regions of plants (i.e., cells and organs differentiate as they grow), and *plant development is brought about by the processes of growth and differentiation.*

## The concept of plant growth hormones

Plant development does not proceed in a random or disorganized manner; on the contrary, there is rather precise regulation of growth and differentiation. To a considerable degree, normal control of plant development is achieved through the agency of extremely small quantities of mobile specific substances, usually called *plant growth hormones*, though sometimes referred to as *growth regulators* or *growth substances*. Etymologically, the word hormone means 'arousing to activity', and it is generally the case that all types of plant growth hormone activate physiological processes that are associated with both growth and differentiation.

An important feature of all plant and animal hormones is that they are synthesized in particular tissues, and are transferred in extremely minute quantities to other regions of the organism where they evoke biochemical, physiological and morphological responses. Growth hormones therefore evoke developmental responses (perhaps,

2

therefore, they should be called 'development hormones', but we will retain the normally used terminology in this book to avoid confusion). In plants, growth hormones are particularly important in the overall coordination and integration of development in different regions of the plant body. They serve as *chemical messengers*, or *signals*, passing from one cell, tissue or organ, to another, and thereby provide a means of communication between different parts of the whole plant. This is in contrast to purely intracellular chemical messengers such as messenger-ribonucleic acid (m-RNA), that operate between organelles within individual cells. Nevertheless, plant growth hormones influence individual cells, and it is clear that their presence can determine the rate and pattern of cellular metabolism; to that extent they operate within as well as between cells.

Although we may recognize what we mean when we speak of plant growth hormones, it is very difficult, if not impossible, to devise a completely unambiguous definition of these substances. The most commonly quoted definition runs along these lines, 'substances which are synthesized in particular cells and which are transferred to other cells where extremely small quantities influence developmental processes'. Now, it can be seen that this definition excludes, for example, sugars, even though these are produced in leaf cells and are translocated to other cells in the plant, where their arrival is marked by either maintenance or speeding up of metabolic activities, which may in turn be reflected in the rate of development. Sugars and other major metabolites, which together provide sources of energy and constructional material, are required in relatively large, and not 'extremely small' quantities.

Greater difficulty is experienced when one considers certain vitamins which are necessary for the activity of many enzymes in plant cells. Thus, a number of the B-group vitamins (nicotinic acid, pyridoxin and thiamin, in particular) are produced in the leaves and translocated to regions such as roots. Normal root development occurs only when adequate, but always very small, quantities of the vitamins are available from the leaves. Consequently, leaf-synthesized B-vitamins could be regarded as growth hormones concerned in the control of root development. A convenient escape from the dilemma of having to classify vitamins as plant growth hormones is provided by knowledge that vitamins are cofactors of known enzymes. Although we cannot rule out the possibility, there is little evidence that plant growth hormones exert their manifold effects by activating

**3**

enzymes in the manner of vitamins. On the contrary, much current research suggests that growth hormones influence the synthesis of enzymes rather than their catalytic activity (p. 153).

Plant growth hormones, then, play a central role in the *internal control* of development, interacting with key metabolic processes such as nucleic acid and protein synthesis. Effects of the *external environment* upon development, too, are apparently often mediated through alterations in hormone concentrations and distribution within the plant (e.g., day length, light quality, or light intensity can influence growth-hormone levels in plants, which in turn leads to changes in the rate or pattern of development).

Plant growth hormones must not merely be present for development to take place; they must also be available in the right place, at the right time, and in proper proportions. This means that (a) the rates of synthesis of growth hormones are carefully controlled, (b) their concentrations in tissues are also regulated by inactivation mechanisms, and (c) the patterns of translocation of growth hormones within the plant are very important. Each of these three aspects of plant growth hormones has consequently received considerable study.

Because plant growth hormones are particularly important in the integration, or correlation, of growth and differentiation in different regions of the plant, they are essentially 'correlation factors', moving from one part of the plant to another part where they exert their effects. Because of this, and their chemically distinct natures, they are often termed *chemical messengers*. It is their particular chemical nature which endows them with their special properties.

In summary, there is great interest in finding out (a) what sort of substances growth hormones are, (b) where in the plant they are manufactured, (c) how their concentrations are regulated, (d) by what means and in what manner they are translocated, (e) what physiological effects they have, and (f) how they exert their effects in cells and tissues. Like many questions, these are more easily asked than answered! Nevertheless, as a result of very active research by many workers all over the world during the last forty or fifty years, we do at least have a general picture of what plant growth hormones exist, where and how they are synthesized, what physiological functions they perform, and their translocation patterns. We are much less sure of the mechanism of growth-hormone action inside individual living cells.

In this chapter we will consider how the currently known plant growth hormones were discovered and chemically identified, what is known of their biochemistry (specifically, how they are synthesized and inactivated inside plants), and how they are translocated. In succeeding chapters the physiological functions of growth hormones are dealt with, and in the last chapter their mechanism of action is considered.

Studies of the chemistry and physiological activities of naturally occurring plant growth hormones have revealed that three major classes of growth-promoting hormones exist in higher plants. These are called *auxins*, *cytokinins* and *gibberellins*. Other classes of hormone may also operate in plants, for example 'growth inhibitors' such as *abscisic acid*. Even a gas, *ethylene*, appears to act as a hormone in a number of physiological processes in plants. In addition to all these, there may be other types of plant hormones of which we are currently unaware, or at least have only indirect evidence to suggest their existence.

Apart from being different chemically, the separate categories of natural plant growth hormones also have distinctive effects in the regulation of plant development. Indeed, we distinguish them on the basis of their physiological roles as much as on grounds of differing chemical constitution. Nevertheless, different classes of plant growth hormones often exert similar developmental responses. Interactions between separate types of growth hormone frequently occur as well, so that the net effect on growth and differentiation of two hormones can be less than, greater than, or completely different from the effect of either one alone.

## Auxins

The auxins are substances produced in small quantities in the apical regions of stems or coleoptiles, in developing seeds, and perhaps in root apices. They are translocated away from their sites of synthesis, and exert effects in other regions, such as stimulating stem or coleoptile elongation, cambial activity, and growth in developing fruit tissues. We shall consider these and other functions of auxins later on.

### The discovery of auxin

Work by Charles Darwin investigating the phenomenon of phototropism (p. 74) in plants, can, in retrospect, be seen to have provided the starting point of research which led to the realization that plants contain growth hormones. Auxin was the first of these to be discovered.

Darwin found that only the apical end, or tip, of the coleoptile in a canary grass (*Phalaris canariensis*) seedling was sensitive to unilateral illumination, but that the growth curvature which caused the coleoptile to bend towards the light source (positive phototropism, p. 74) occurred some distance below the tip in the elongating portion of the coleoptile (Fig. 1.1). That is, the tip *senses* the unilateral light stimulus, but the curvature *response* takes place lower down as a

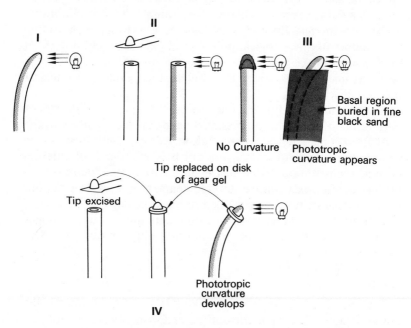

*Figure 1.1.* Some early experiments which suggested the existence of a growth hormone in plants. I. positive phototropic curvature by coleoptile of a grass or cereal seedling. II. Darwin's experiment of 1880 which demonstrated that the apical part of a coleoptile must be present for a phototropic response to occur. III. Darwin's observations which revealed that perception of unilateral illumination takes place in the coleoptile tip, and yet curvature occurs lower down the organ. IV. an experiment by Boysen-Jensen in 1911, which showed that the growth curvature-inducing factor from the tip was able to pass through agar-gel, or gelatine, and was hence presumably purely chemical in nature.

result of more rapid growth on the shaded than the illuminated side of the coleoptile. Darwin also noted that when the extreme tip of the coleoptile was excised no curvature response occurred in the remaining part of the organ. He concluded in his book, *The Power of Movement in Plants*, published in 1881, that 'when seedlings are freely exposed to a lateral light some influence is transmitted from the upper to the lower part, causing the latter to bend'. The important point to note here is that Darwin conceived the idea of a growth correlation factor, i.e., 'influence', which is transmitted from a coleoptile tip to the lower elongating region. He did not, of course, have any idea of the nature of the correlation factor. Its existence was, however, established beyond doubt in experiments conducted by several workers in the first two decades of this century, and it was also demonstrated to be of a purely chemical nature when it was revealed that the 'influence' could pass through a layer of gelatine placed between an excised coleoptile tip and the lower responding region (Fig. 1.1). These early investigations were mainly concerned with the phenomenon of phototropism in plants (see p. 74). It was subsequently suggested that the growth curvature-inducing influence which originates in a unilaterally lit coleoptile tip, also serves as a correlative growth promoter (i.e., as a growth hormone), in normal straight elongation growth.

Not until 1926 was this growth hormone clearly isolated from plant tissues. In that year a Dutchman, F. W. Went, reported that when an excised oat (*Avena*) coleoptile tip was left on the surface of a small block of agar-gel for a few hours, then the agar block acquired the property of promoting growth in other coleoptiles without a tip being present (Fig. 1.2). Thus, the chemical influence, first proposed by Darwin, was now shown to diffuse out of a detached coleoptile tip into an inert aqueous medium such as agar-gel, and to retain for some time its capacity for inducing growth.

Went measured the quantity of the growth-promoting factor obtained in agar blocks by placing each block on one side of a decapitated oat coleoptile (i.e., a coleoptile from which the tip had been cut off), and subsequently determining the extent of the growth curvature which developed in darkness (Fig. 1.2). The greater the angle of bending in the coleoptile then the greater was the elongation growth of the coleoptile in the region situated below the asymmetrically positioned agar block. It was found that higher concentrations of the growth hormone in agar produced more curvature than

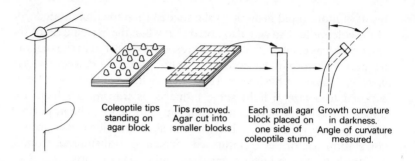

Coleoptile tips
standing on
agar block

Tips removed.
Agar cut into
smaller blocks

Each small agar
block placed on
one side of
coleoptile stump

Growth curvature
in darkness.
Angle of curvature
measured.

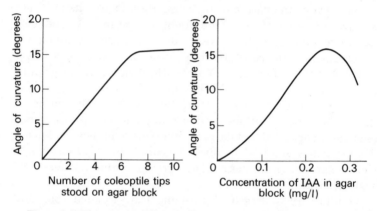

*Figure 1.2.* The way in which auxin from coleoptile tips was first clearly isolated from living tissues by F. W. Went in 1928. Auxin diffused from severed tips into agar. The agar thus acquired the capacity to induce elongation growth in coleoptiles from which the tip had been removed. By placing an agar block containing auxin on to one side of a decapitated coleoptile, a growth curvature developed below. Measurement of the angle of curvature afforded a means of estimating the auxin content of the block. This was the first bioassay devised for auxin measurement, and two dose-response curves are shown; one for unknown auxin from coleoptile tips, and the other for indole-3-acetic acid (IAA) included in agar blocks.

(Adapted from P. F. Wareing and I. D. J. Phillips, *The Control of Growth and Differentiation in Plants*, Pergamon Press, Oxford, 1970.)

lower concentrations, and consequently Went had devised the first means of quantitatively measuring a plant growth hormone. The technique did, of course, depend on a biological response to applied hormone, and is therefore an example of a biological assay, or *bioassay*, of which a great many are now known for measuring the various types of plant growth hormones. The growth hormone isolated from oat coleoptile tips was given the name of *auxin* (from the Greek *auxein*, to grow), and it has subsequently become clear that auxin occurs in all higher plants.

## The chemistry of auxins

We generally speak of auxins, rather than the singular, auxin. Early workers thought that they were dealing with only one substance in their experiments with coleoptiles, but later studies have revealed that a number of substances occur in plants which possess activity in bioassays similar to that shown by the auxin obtained from coleoptile tips. The first auxin identified chemically was indole-3-acetic acid (often abbreviated to IAA) (Fig. 1.3) following its isolation from

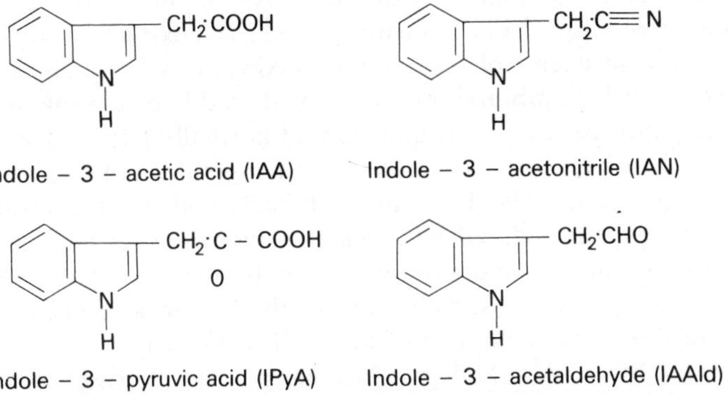

Indole – 3 – acetic acid (IAA)   Indole – 3 – acetonitrile (IAN)

Indole – 3 – pyruvic acid (IPyA)   Indole – 3 – acetaldehyde (IAAld)

*Figure 1.3.* Some naturally occurring indole compounds in plants, all of which may show auxin-like activity when applied to plants. It is probable that only IAA serves as an auxin, and that other indoles, such as those depicted, are metabolized in plant tissues to yield IAA.

human urine. In the following year, 1935, the auxin present in cultures of the mold *Rhizopus suinus*, and in yeast suspensions, was also found to be IAA. The reason for chemists' attention being devoted to these sources was that early attempts to isolate and chemically characterize auxins were thwarted by the extremely low concentrations in which growth hormones occur in higher plants. Insufficient quantities of auxin were available for the preparation of crystalline material required for chemical analysis.

In the years since 1935, the majority of plant physiologists have come to believe that IAA is the principal auxin present in all plant species. Nevertheless, other chemicals have been isolated from plants which act as auxins (Fig. 1.4), and some of these are indole compounds closely related chemically to IAA. For example, indole-3-acetonitrile (IAN), indole-3-pyruvic acid and indole-3-acetaldehyde occur in many species (Fig. 1.3). It seems likely that naturally occurring indoles other than IAA may be precursors of IAA in a biosynthetic sequence. Evidence in support of this view is seen in that, for example, IAN promotes growth only in those plant species which contain the enzyme required to convert the compound to IAA.

Auxins are obtained from plant tissues in various ways. We saw earlier that the first isolation of an auxin was achieved by allowing the hormone to diffuse out from the cut surface of a plant organ into a suitable inert medium such as agar-gel. Auxins may also be obtained from plant materials by extraction. An organic solvent, such as methyl or ethyl alcohol, is usually used as an extraction solvent due to the high degree of solubility of auxins in such liquids, and also because cell membranes are penetrated readily by organic solvents. The plant tissues are ground up and normally left to steep in the organic solvent for some hours at a temperature of 0°C to reduce enzyme activity. The tissues are then filtered off, and the filtrate contains a mixture of dissolved substances, including auxins. The filtrate has to be purified before the auxins can be measured, either in a bioassay or by physical and chemical methods. Purification can be performed by use of a number of standard chemical techniques, such as paper-, thin layer-, and gas-chromatography, electrophoresis, and ion-exchange systems. All of these separation procedures distribute the constituents of the original plant extract between a number of separate fractions, each of which can then be tested for auxin activity in some form of bioassay (Fig. 1.4). Auxins obtained from plants by extraction are often referred to as *extractable auxins*, in contrast to *diffusible auxins* which are collected in agar-gel from the cut surface of plant tissues. There are sometimes qualitative and quantitative differences between extractable and diffusible auxins, even when they are isolated from the same plant tissue, but the full significance of such differences is not yet clear to us. Active fractions can be studied further in an attempt to identify the substances responsible for auxin activity, but this phase of an investigation is always much more difficult than the relatively simple demonstration of the existence of activity in a bioassay.

These general methods of extracting, purifying, measuring and

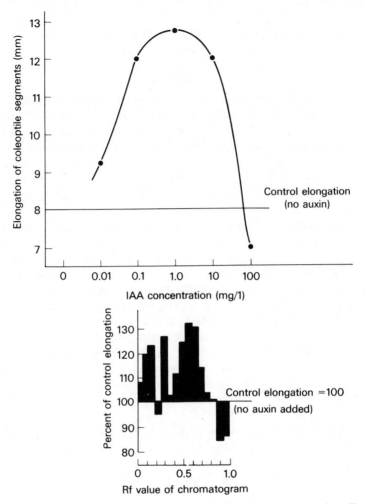

*Figure 1.4. Above.* Dose-response curve for IAA in the wheat coleoptile segment straight-growth assay. Etiolated coleoptile segments were originally 10 mm long, and elongated by the indicated amounts within 20 hr at 25°C in darkness. Note that maximum elongation occurred with 1.0 mg/l IAA ($5.7 \times 10^{-5}$ M IAA), and that concentrations higher than this are supra-optimal. Growth inhibition by higher auxin concentrations appears to be a result of enhanced ethylene biosynthesis. *Below.* A typical result of a coleoptile straight-growth assay of a chromatographed plant extract. In this example, developing peach embryos were extracted. Note that more than one 'peak' of auxin activity is revealed, which suggests that plants contain more than just one type of auxin.

(From G. M. Weaver and L. F. Hough, *Amer. J. Bot.* **46**, 718–724, 1959.)

identifying auxins are also used in studies of other classes of plant growth hormones, such as gibberellins and cytokinins.

## Indole-3-acetic acid (auxin) metabolism

Studies of the metabolism of naturally occurring auxins have been concentrated almost exclusively on indole-3-acetic acid (IAA), due to the strong evidence which suggests that this substance serves as the principal auxin in most, if not all plants. It has been found that IAA concentrations in plants are controlled by variations in, (a) the rate of auxin synthesis, (b) the rate of auxin destruction, and (c) rates of auxin inactivation by means other than actual degradation of the IAA molecule.

*Auxin biosynthesis.* This occurs principally in the shoot-tip region, particularly in the young expanding leaves of the apical bud, or in the tip of a grass or cereal seedling coleoptile, and in developing embryos. We know that the rate of auxin synthesis varies, and is influenced both by the external environment and physiological age of the plant or individual organ. Thus, for example, in green tissues auxin synthesis is greater in the light than in darkness, and greater quantities of auxin are produced in developing leaves and fruits at certain stages of their development than at others (pp. 58 and 102). Auxin synthesis in perennial plants also changes with the seasons, so that in the twigs of woody species for example, auxin levels are high in the spring and early summer but very low during the autumn and winter months.

The biosynthetic pathways of IAA formation are not yet fully understood. Available evidence suggests that several alternative routes of IAA synthesis exist in plants, all starting from the amino acid tryptophan (Fig. 1.5). Thus, when tryptophan is supplied to most plant tissues, IAA is formed. Some zinc-deficient plants, the stems of which do not elongate, can be induced to elongate normally by the addition of either an auxin or tryptophan. This suggests that tryptophan is lacking in such zinc-deficient plants, and that growth is checked because of the absence of this precursor of IAA. However, recent research with certain species, particularly oat, has shown that exogenous tryptophan is converted to IAA only when the plant tissues are contaminated with bacteria. Under sterile conditions, oat tissues are unable to synthesize IAA from tryptophan. There is, therefore, some doubt at present as to whether tryptophan serves as a precursor of IAA in all species of higher plants.

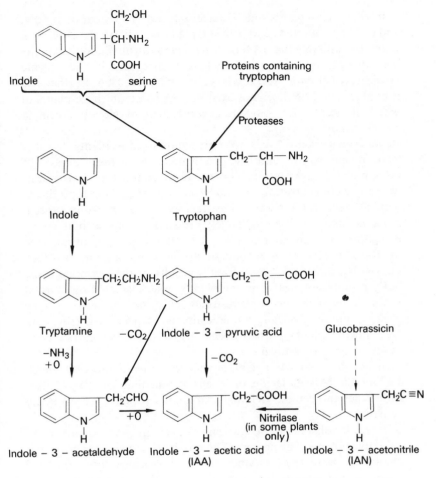

*Figure 1.5t* Possible routes in the biosynthesis of indole-3-acetic acid (IAA) in plant tissues.

Enzymic transformations from tryptophan to indole-3-pyruvic acid, to indole-3-acetaldehyde and thence to IAA, all of which certainly occur in bacteria but may also operate in some plants, are shown in Fig. 1.5. In a few species, including oat, the pathway through tryptamine to IAA (Fig. 1.5) also appears to be functional, but it is not known whether tryptamine is normally formed from tryptophan, or from indole.

In a number of plants, indole-3-acetonitrile (IAN) occurs naturally, and can be converted to IAA in some, but not all, species (Fig. 1.5). It appears unlikely that IAN is itself active as an auxin, for it exhibits activity only in those species whose tissues contain the enzyme nitrilase, which catalyses the conversion of IAN to IAA. Further, it is probable that IAN does not exist in a free state in those plants in which it occurs, but only as a constituent of glucobrassicin, a thioglucoside.

**Auxin inactivation.** IAA is not only synthesized in plants, but it is also, in one way or another, inactivated during the processes of growth and differentiation. This has been demonstrated in various ways. The concentration of *endogenous auxin* (i.e., auxin produced within the plant) is highest in those regions where it is manufactured, remains almost as high in actively growing regions such as young elongating internodes, but falls off to very low levels in matured tissues (Fig. 2.5). Also, introduction of IAA into a plant (it is then said to be *exogenous auxin*) usually results in the hormone being fairly rapidly inactivated in the plant's tissues, so that the normal auxin status of the tissues is restored. This does not however take place so rapidly if certain synthetic auxins, such as 2,4-D (Fig. 6.3), are applied to plants, probably because of a lack of enzymes suited to catalyse their breakdown.

Inactivation of IAA in plants is achieved either by processes which degrade the IAA molecule, or by enzyme-mediated linking of IAA with other molecules, yielding compounds which have no auxin activity.

The catabolism, or breakdown, of IAA in plant tissues is an oxidative process. The process of IAA oxidation does not necessarily require the presence of enzymes, for rapid photo-oxidation (i.e., the energy of light can induce oxidation) of IAA will take place in a solution exposed to light. Some pigments present in plants, particularly riboflavin and violaxanthin, may well absorb light energy which can be used to energize the oxidation of IAA (see also p. 79). It is probable, however, that photo-oxidation is of less importance in the regulation of endogenous IAA levels than is non-light-requiring enzyme catalysed oxidation. Plants contain an enzyme or enzyme system, known as *IAA-oxidase*, which catalyses the breakdown of IAA. The exact nature of this enzyme has yet to be established, but it may play a role in controlling endogenous IAA levels. This is suggested by only circumstantial evidence to date, in that it has been found by several research groups that regions of the plant low in auxin have at the same time a higher level of IAA-oxidase activity than auxin-rich tissues (Fig. 1.6). However, results of experiments comparing IAA-oxidase levels in different tissues must be treated with caution, for it is known from work by Galston and his associates that IAA-oxidase activity may be altered by the presence of certain naturally occurring cofactors and inhibitors of the enzyme.

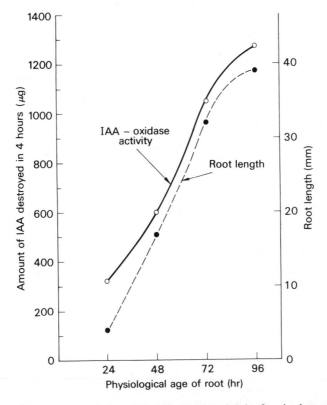

*Figure 1.6*. Older tissues appear to possess higher levels of auxin-destroying enzymes than young growing tissues. In the example illustrated, enzyme preparations were obtained from 2 g fresh weight of *Lens culinaris* roots of various ages, and incubated with 1750 μg IAA for 4 hr. The greater the proportion of old tissues used for enzyme preparation, the greater was the IAA-oxidase activity.

(Adapted from T. Gaspar, A. Xhaufflaire, and M. Bauillene-Walrand, *Bull. Soc. Sciences Liège* **34**, 149–175, 1965.)

The metabolic products of IAA breakdown in plants have not yet been fully identified, but they include indolealdehyde and compounds such as 3-methylene-2-oxindole (Fig. 1.7).

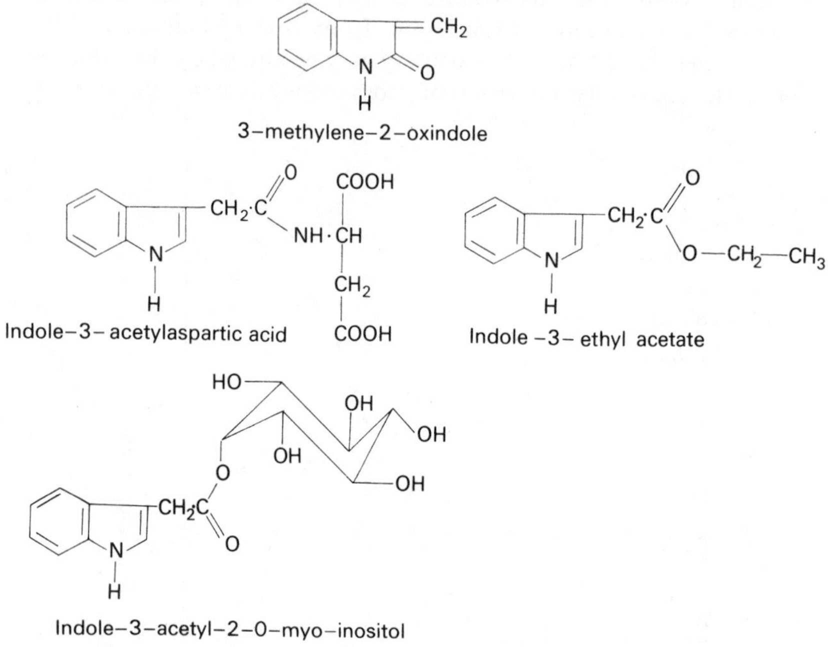

3−methylene−2−oxindole

Indole−3−acetylaspartic acid

Indole −3− ethyl acetate

Indole−3−acetyl−2−0−myo−inositol

*Figure 1.7. Top.* 3-methylene-2-oxindole, which is a product of the action of IAA-oxidase upon IAA, and does not possess auxin activity. *Centre and Bottom.* indole-3-acetylaspartic acid is an inactive conjugate of IAA and aspartic acid. The esters, indole-3-ethyl acetate and indole-3-acetyl-2-O-myo-inositol both retain the capacity to elicit typical auxin responses.

In addition to oxidation processes which break down IAA, plants possess mechanisms whereby IAA can be joined with other molecules to yield conjugates which may retain hormone activity, or be physiologically inert. Thus, plant enzymes act on exogenous IAA to yield hormonally inactive peptides with aspartic or glutamic acid (e.g., indole-3-acetylaspartic acid, Fig. 1.7). Alternatively, esters may be formed between IAA and sugars (e.g., the glucosyl ester of IAA), or between IAA and a sugar alcohol, myo-inositol (indole-3-acetyl-2-0-myo-inositol, Fig. 1.7), which retain hormone activity when supplied to plant tissues. There is also some evidence that exogenous

IAA can be adsorbed onto plant protein, and that the ester, indole-3-ethyl acetate (Fig. 1.7), can occur naturally in plants.

**Auxin translocation**

It is normally the case that the site of auxin production and its place of action in the plant body are separated by some finite distance. Thus, to reach its destination, auxin is translocated. Studies of auxin translocation have shown that this class of plant growth hormone is transported much more readily in one direction in plant tissues than in the opposite direction. This peculiar feature of auxins was first discovered by Went in 1928, who followed the movement of auxin in etiolated oat coleoptile segments. Agar blocks containing auxin ('donor blocks') derived from coleoptile tips (probably IAA) were placed on either the morphological upper ends of the segments, or on the morphological lower ends. 'Receiver blocks' of pure agar alone were placed on the opposite ends (Fig. 1.8). The auxin moved out from the donor blocks into and through the coleoptile, and eventually emerged into the receiver blocks. The quantity of auxin appearing in the receiver blocks was determined in the *Avena* curvature bioassay (Fig. 1.2). Went found that regardless of the orientation of the coleoptile segments with respect to gravity, auxin was obtained only in receiver blocks placed at the morphological basal end. In other words, auxin moved only from the morphological upper to the lower end of the coleoptile, revealing the existence of a *basipetal* (i.e., from apical to basal end) *polar transport* mechanism. *Acropetal transport* (from base to apex) of auxin did not appear to take place at all in *Avena* coleoptile segments.

Since Went's experiments with coleoptiles were performed, basipetal polarized auxin transport has been found to take place in plant organs such as stems, hypocotyls, and petioles of dicotyledonous plants. Many of the more recent experiments investigating this phenomenon have used carbon-14 labeled IAA ($^{14}$C-IAA), the movement of which can be followed without recourse to a bioassay (Fig. 1.9). It is now known that auxin transport is not always as strictly polar as Went found. Some acropetal movement of auxin occurs in all plant organs (Fig. 1.9), but in shoot tissues it is very much slower than basipetal transport, and appears to be the result of nothing more than physical diffusion of auxin molecules within the continuous aqueous phase of plant tissues. This has been established in experiments which have shown that in contrast to acropetal transport,

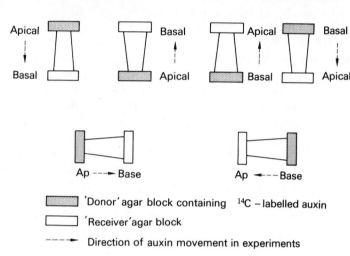

**Measuring basipetal transport**  **Measuring acropetal transport**

Apical

Basal

Basal

Apical

Apical

Basal

Basal

Apical

Ap ---► Base        Ap ◄--- Base

▨▨▨ 'Donor' agar block containing $^{14}$C – labelled auxin

☐ 'Receiver' agar block

---► Direction of auxin movement in experiments

*Figure 1.8.* The agar *donor–receiver* block method used in studies of transport of plant growth hormones, particularly auxins. Most studies have utilized $^{14}$C-labeled auxins applied in the donor block to one end of stem, petiole, coleoptile or root segments. In the diagram, the narrower end of a segment indicates that that is the morphological more apical end. Basipetal and acropetal movement of the hormone is followed with segments either in their normal, preferred, orientation, or inverted, or horizontally positioned.
(Adapted from M. H. M. Goldsmith, chapter 4 in *The Physiology of Plant Growth and Development* (ed. M. B. Wilkins), McGraw-Hill, London, 1969.)

basipetal polar auxin transport involves the consumption of metabolically derived energy, and can therefore be inhibited by lack of oxygen or by treatment with metabolic inhibitors (Fig. 1.10). Since acropetal auxin movement is a physical process, it goes on even under anaerobic conditions.

The actual *velocity* (distance moved in unit time) of basipetal polar auxin transport is quite low, having been found to range from 5 to 15 mm/hr at temperatures of 20 to 25°C in many aerial plant organs (Fig. 1.11). This rate is nevertheless much greater than the rate of acropetal auxin transport in the same organs.

For many years there has been some confusion as to the direction of polar auxin movement in roots. Recent research has, however, clearly demonstrated that movement of IAA in roots is definitely polar but *acropetal*, i.e., movement of auxin takes place more rapidly toward than away from the root tip (Fig. 1.12). The full physiological significance of this situation in roots is not yet clear, but is referred to again during considerations of growth and growth movements in roots (pp. 68 and 88).

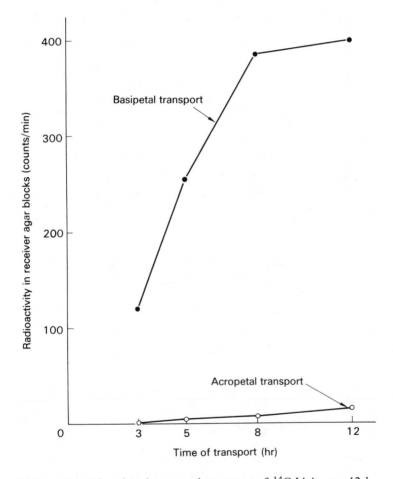

*Figure 1.9.* Basipetal and acropetal transport of $^{14}$C-IAA over 12 hr through petiole segments from *Phaseolus vulgaris*, measured by the agar donor–receiver block method (see Fig. 1.8).
(From C. C. McCready and W. P. Jacobs, *New Phytol.* **62**, 19–34, 1963.)

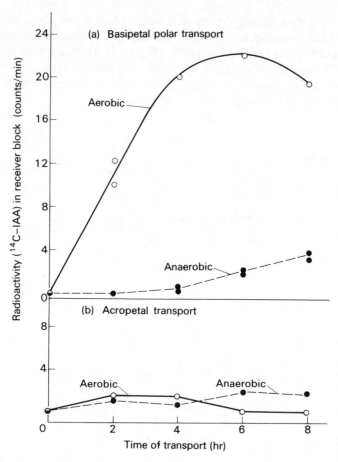

*Figure 1.10.* Effect of anaerobic conditions upon polar basipetal, and nonpolar acropetal, transport of auxin in oat coleoptile segments. The agar donor–receiver block method was used (see Fig. 1.8), with $^{14}$C-IAA applied to 5 mm long segments. Lack of oxygen depressed polar basipetal transport of IAA (a), but had no significant effect on the already very low velocity of acropetal transport (b).

(From M. B. Wilkins and M. Martin, *Plant Physiol.* **42**, 831–839, 1967.)

The actual tissues which conduct auxin have not yet been determined with any degree of certainty. In the case of coleoptiles, the available evidence suggests that all cells are able to transport auxin basipetally at approximately 1 cm/hr. Basipetal movement of auxin from the young leaves of the apical bud in dicotyledonous plants might not be expected to occur in either the phloem or xylem, for movement of solutes in the vascular tissues of these regions is upward, carrying photosynthate, water and inorganic ions into the young developing leaves. Also, known velocities of solute movement in phloem and xylem are greatly in excess of 1 cm/hr.

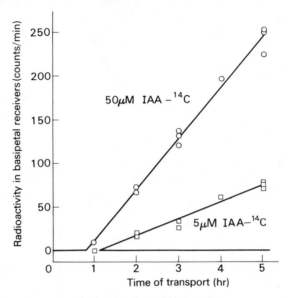

*Figure 1.11*. Estimation of the velocity of basipetal auxin transport ([14]C-labeled IAA) in *Phaseolus vulgaris* petiole segments. Two concentrations of IAA in agar were used (5 and 50 $\mu$M). Calculated intercepts of 1.12 and 0.80 hr for the 5 and 50 $\mu$M IAA were not significantly different from one another. Since the petiole segments were 5.44 mm long, the mean calculated velocity of IAA transport was 5.44 mm per 0.96 hr, or 5.7 mm/hr, for polar basipetal transport.
(From C. C. McCready and W. P. Jacobs, *New Phytol.* **62**, 19–34, 1963.)

***Morphological indications of polar auxin transport.*** The polar transport of auxin is undoubtedly of great importance in the control and coordination of growth and differentiation in different parts of a plant. A number of polar phenomena in plants are correlated with polarity of auxin transport. These provide support for the view that the pattern of movement of endogenous auxin is accurately reflected in the results of the numerous experiments conducted with isolated segments of plant organs and exogenous auxins.

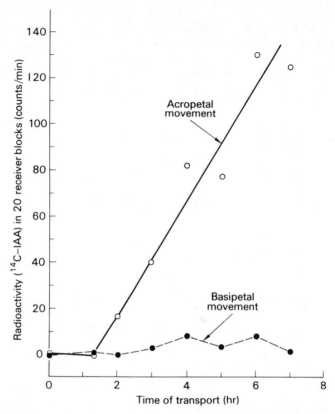

*Figure 1.12.* Polar auxin transport in root segments is acropetal. The data shows transport of $^{14}$C-IAA in 6 mm long segments of *Zea mays* roots, following its application in agar at a concentration of 1 $\mu$M.
(From T. K. Scott and M. B. Wilkins, *Planta* **83**, 323–334, 1968.)

In deciduous woody species, the vascular cambium (p. 55) is first stimulated to activity in springtime immediately below the young expanding buds which are centers of auxin synthesis. A wave of cambial activity then spreads basipetally through twigs, branches and down the trunk. This accords with polar basipetal auxin transport in these regions, and the known stimulatory effect of auxin on cambial activity (p. 56). In the root system of trees, cambial activity commences when the wave of renewed cambial activity reaches the base of the trunk and continues acropetally into and along the roots. In roots as in stems, therefore, there is a correlation between the polarity of auxin transport and direction of movement of the cambial stimulus.

In phototropic responses in coleoptiles, the growth curvature (p. 73) first develops just below the region of auxin synthesis in the tip, and then moves basipetally just as auxin does in the same organ.

A number of regeneration phenomena in plants can be related to polar transport of auxin. For example, initiation of roots in shoot cuttings is promoted by auxins, and new roots are normally formed at the morphologically basal end of the stem due to the arrival there of basipetally transported endogenous auxin. Root regeneration in segments of roots, such as those of convolvulus, dandelion or dock, on the other hand, occurs at the morphologically apical end of the root segment, and this may be a reflection of acropetal auxin transport in these organs.

## Gibberellins

The class of plant growth hormones which we now call the gibberellins consists of a number of compounds which exhibit growth-promoting properties that are in some ways similar to, but in others different from those of auxins. The chemical nature of the gibberellins is distinct from all known auxins. The gibberellins are, however, very closely related to one another chemically (Fig. 1.13). One or more of the gibberellins have been found to be present in plants of all angiosperm species examined, and also in some gymnosperms, pteridophytes and algae. It seems safe to say that all plants, certainly all higher plants, contain gibberellins and that these substances serve as natural growth hormones. This was not, however, realized at the time of their discovery.

The gibberellins were really discovered by accident. At least, their discovery was not a consequence of a deliberate study of plant growth hormones. In fact, the story is a fascinating one and something of an object lesson in the way in which observations on one field of enquiry can lead to a revolution of thought in another area of science. Unfortunately, it is also an example of the delay in progress which can occur as a result of the intrusion of war into the processes of scientific research. The story started just before the start of the twentieth century, when Japanese farmers noticed that some of their paddy fields contained unusually tall rice seedlings. At first the farmers were probably excited at the prospect of a better crop than usual, and felt that they had been blessed with a new vigorous strain of rice. In this sense the gibberellin story has a happy beginning.

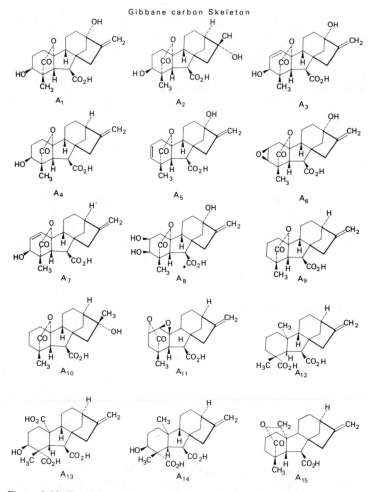

Gibbane carbon Skeleton

*Figure 1.13. Top.* The 'gibbane carbon skeleton', which is common to all known gibberellins. *Below.* Fifteen of the currently known gibberellins, $A_1$, $A_2$, etc., some of which were isolated from the fungus, *Gibberella fujikuroi*, and others from higher plants. $A_3$ = gibberellic acid.
(From B. D. Cavell, J. MacMillan, R. J. Pryce, and A. C. Sheppard, *Phytochemistry* **6**, 867–874, 1967.)

Disappointment was soon felt though, when the abnormally tall seedlings failed to live to maturity and often fell over into the water and mud. It was soon realized that the excessive elongation growth in the stems and leaves of the seedlings was caused by an infection of some sort. The syndrome was quickly called the *Bakanae*, or 'foolish seedling', disease. In 1926, Kurosawa, a plant pathologist working in Taiwan, discovered that 'foolish' rice seedlings were infected with the fungus *Gibberella fujikuroi* (also known as *Fusarium moniliforme*). A number of Japanese workers followed up this finding, particularly Yabuta and Sumiki, and discovered that when the fungus was grown in a culture medium, it secreted something into the medium which if applied to rice plants would induce the same growth response as the fungus itself. By 1939, the Japanese had isolated from *Gibberella fujikuroi* a small quantity of highly active crystalline material which was named *gibberellin A*. At that time its chemical nature was not known, but Japanese workers made steady progress during the war years towards its identification. Western scientists did not become aware of the Japanese work until approximately ten years later, largely because of the second world war.

In 1954 a major advance was made, when work which had been prompted by the earlier Japanese results reached fruition. In that year, British chemists isolated and chemically characterized a pure growth promoting compound from *Gibberella fujikuroi*. It was found to have the structure shown in Fig. 1.13, and was named *gibberellic acid*. Quantities of gibberellic acid were made available by Imperial Chemical Industries to research workers all over the world, who discovered very quickly that many species of plant would, when treated with the substance, react by growing taller than usual. However, it was realized that the nature of the responses shown were essentially *normal* in character, in that distorted growth patterns did not occur. Because of this, it was suspected that perhaps gibberellic acid was also a natural, endogenous, growth hormone in higher plants. Extracts of a number of healthy noninfected plants were very soon found to contain substances possessing biological activity and chemical constitution similar to those of gibberellic acid. All these, together with gibberellic acid, were called *gibberellins*. The gibberellins are based chemically upon what is called the *gibbane carbon skeleton* (i.e., the basic framework of all gibberellins; Fig. 1.13). The known gibberellins have been assigned numbers, always with the prefix 'A'. Thus we have gibberellin $A_1$, gibberellin

$A_2$ and so on, with gibberellic acid designated as gibberellin $A_3$. They are often abbreviated to $GA_1$, $GA_2$, $GA_3$, etc. Some of the twenty-eight currently known gibberellins have been isolated from culture filtrates of *Gibberella fujikuroi*, and others from various organs of a range of species of higher plants. The methods used in obtaining gibberellins from plants are virtually identical to those employed in studies of auxins (p. 10).

In the mid-1950's, as a consequence of the discovery of the gibberellins and also cytokinins (see p. 30). it became apparent that auxins were not the only growth hormones involved in the normal control of plant development. Plant physiologists had to learn to think in terms of several categories of growth hormone, each with its independent effects but also interacting with one another in a number of growth phenomena, as we shall see in later chapters.

### Gibberellin metabolism

Endogenous gibberellins appear to be synthesized in the same regions of the plant as auxins (p. 12), but not necessarily at the same time nor at identically varying rates. Thus, young leaves (Fig. 1.14), developing embryos (p. 103) and root apices (Fig. 1.14) are known to be centers of gibberellin production. The environment influences gibberellin synthesis so that, for example, plants grown under long days usually produce more gibberellins than those under short days. Also, the stage of development of an organ may be a determining factor, and this appears to be the case in embryos which produce gibberellins more actively at some phases of their development than at others.

We know quite a lot about the biochemical pathways of gibberellin synthesis, but very little of the catabolic processes which bring about gibberellin inactivation in plants.

*Gibberellin biosynthesis.* No gibberellins have yet been made *in vitro* (i.e., outside a living organism). Their chemical nature is such as to make this extremely difficult. This has nevertheless not prevented biochemists elucidating the route by which enzyme-mediated gibberellin synthesis occurs *in vivo* (i.e., within living plants).

The early stages of gibberellin biosynthesis are the same as for the synthesis of the many compounds classified as terpenoids, of which a great number occur in plants (e.g., carotenoids, sterols and the phytol moiety of chlorophyll). All terpenoids are built up from a number of *isoprene units*, each unit containing five carbon atoms. Two isoprene

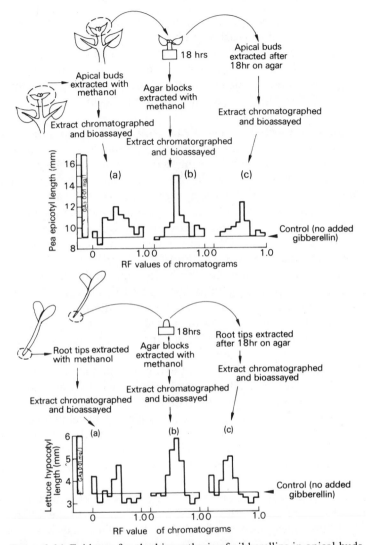

*Figure 1.14.* Evidence for the biosynthesis of gibberellins in apical buds and root tips of sunflower plants. Note that greater quantities of gibberellin activity were obtained by diffusion into agar (b) than by extraction of the tissues before (a) or after (c) 18 hr on agar. Apical bud extract chromatograms were assayed with the pea epicotyl elongation test, and those of the root tips with the lettuce hypocotyl elongation test.

(From R. L. Jones and I. D. J. Phillips, *Plant Physiol.* **41**, 1381–1386, 1966.)

units joined together make a monoterpene (ten carbon atoms), three a sesquiterpene (fifteen carbon atoms), four a diterpene (twenty carbon atoms), etc. Gibberellins are all closely related to the diterpenes, containing either twenty or nineteen carbon atoms, and the sequence of enzymic reactions which leads to the formation of a gibberellin is similar to that for all diterpenes up until the final stages of its formation (Fig. 1.15).

Studies of gibberellin biosynthetic pathways have been aided by the discovery that certain commercially available 'growth-retardants' can selectively inhibit specific steps in gibberellin biosynthesis. These compounds are known as 'AMO' (2'-isopropyl-4'-(trimethylammonium chloride)-5'-methylphenyl-1-piperidine carboxylate), and 'Cycocel', or 'CCC', (2-chloroethyltrimethylammonium chloride). The steps in gibberellin biosynthesis which are selectively blocked by AMO and CCC are indicated in Fig. 1.15. A third growth retardant known as 'Phosfon-D' (tributyl-2,4-dichlorobenzylphosphonium chloride) also inhibits gibberellin biosynthesis, but rather less specifically than AMO or CCC since in addition it interferes with the biosynthesis of other cyclic diterpenes (Fig. 1.15). It is highly likely that the retarding effect on plant growth of these substances is a consequence of their capacity to prevent gibberellin biosynthesis in treated plants.

*Gibberellin inactivation.* In contrast to the clear picture of gibberellin biosynthesis which is available, we have little idea of the mechanisms of gibberellin inactivation in plants. As with auxins, there is evidence that gibberellins are bound in some way to cellular constituents which renders them inactive as hormones. We do not know the nature of such conjugates, but active gibberellins can be released from such bound forms under hydrolysing conditions. Plants do however contain enzymes which quite readily attack a gibberellin molecule, but usually an applied gibberellin is merely converted to another gibberellin. In other words, interconversion of the different gibberellins takes place in plants (Fig. 1.15). This may have considerable physiological significance, but a great deal of work remains to be done to establish what it is.

### Gibberellin translocation

In contrast to the polar nature of auxin transport in plants (p. 17), gibberellins move readily in all directions within the plant. This has been demonstrated by the use of radioactive gibberellic acid, which spreads throughout the plant regardless of the point chosen for its application (Fig. 1.16). The movement goes on in all tissues, including phloem and xylem, again contrasting with the movement of auxins (p. 20).

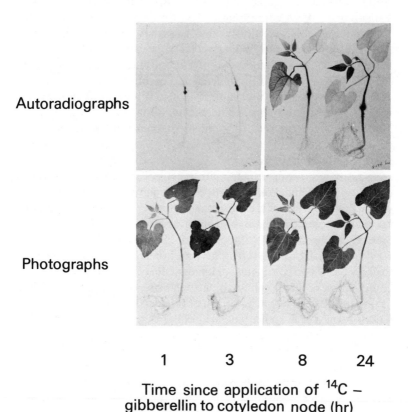

*Figure 1.16.* Nonpolar movement of gibberellin in bean plants occurs in all tissues, including xylem and phloem. $^{14}C$-labeled gibberellin was applied to the cotyledonary node, and its movement traced by autoradiography at intervals of time.

(From G. Zweig, S. Yamaguchi, and G. W. Mason, *Advances in Chemistry Series No. 28*, 122–134, 1961. Original print supplied by Dr. Gunter Zweig.)

## Cytokinins

Cell-division activity in isolated plant tissues, or young embryos, normally requires the supply of a cell-division-inducing factor in addition to the provision of all necessary nutrients. Thus, for example, van Overbeek in 1941 found that very young excised plant embryos would continue development in sterile culture only when nurtured in a nutrient medium containing coconut milk (a liquid endosperm). The coconut milk was not needed to satisfy the gross nutritional demands of the embryos, but was effective because it contained very low concentrations of unknown substances which allowed continued cell division activity within the cultured embryos.

For many years coconut milk, and also other liquid endosperms such as that from immature fruits of horsechestnut (*Aesculus hippocastanum*), has been used as a supplement to nutrient media in studies of isolated plant embryos, organs, tissues and cells. Particularly rigorous studies have been made of the growth requirements of cultures prepared from parenchymatous cells of the tobacco (*Nicotiana tabacum*) stem, and from the phloem parenchyma of carrot (*Daucus carota*) roots. Much of the important work in this field has been done by two independent research groups, one at Cornell University led by F. C. Steward and the other at the University of Wisconsin under F. Skoog. Both of these researchers and their co-workers attempted to grow tissues on completely defined media (i.e., media the chemical composition of which is known), and this presented many problems. Skoog found at first that a portion of tobacco stem pith parenchyma would form a true *callus* (i.e., an undifferentiated mass of dividing cells) only when given a nutrient medium containing coconut milk. Later, in the mid 1950's, he made the most important discovery that coconut milk could be dispensed with if the nutrient medium contained an auxin and also the purine base *adenine*. If auxin (IAA) was supplied in the absence of adenine, then the cells of the isolated pith tissue enlarged a little but did not divide. Adenine given without auxin had no effect other than to slightly enhance DNA synthesis in the cells. Thus both auxin and adenine were required for the initiation of cell division activity in a callus (Fig. 1.17).

Adenine is a purine derivative and a component of nucleic acids (Fig. 1.18). Skoog and Miller found that DNA preparations from plants or animals were even more active than adenine in promoting cell-division activity in plant cells, *provided that the DNA was broken down by heating.* Considerable work by Skoog, Strong, and Miller led to the isolation of the active factor from broken-down DNA. It was identified as 6-(furfurylamino) purine, and is therefore similar to adenine in that it is a purine derivative. Because of its great power to stimulate cell division (cytokinesis) in the presence of an auxin, 6-(furfurylamino) purine was given the name *kinetin* (Fig. 1.18).

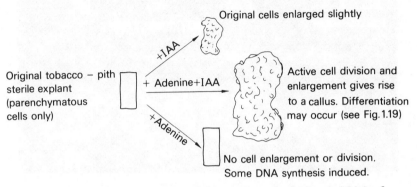

*Figure 1.17.* Diagram to illustrate the response to adenine and IAA of excised parenchymatous tissue in sterile culture.

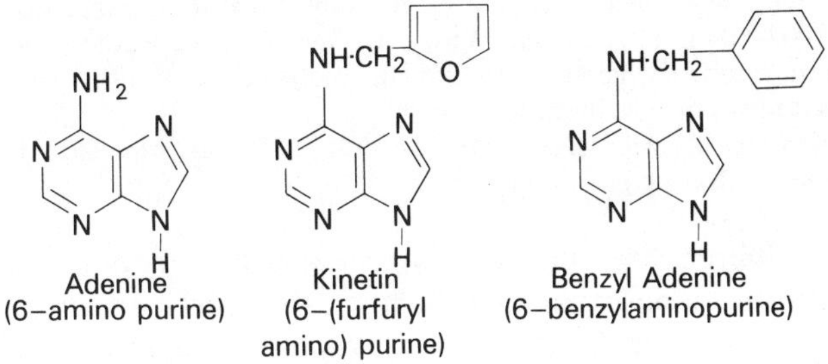

*Figure 1.18.* Two synthetic cytokinins, kinetin and benzyl adenine, illustrating their chemical relationship to the naturally occurring purine base, adenine.

In addition to inducing cell division in isolated plant tissues, IAA and kinetin were found to cause the onset of differentiation in a previously undifferentiated callus. Further, particular patterns of differentiation could be obtained by simply varying the proportions of IAA and kinetin in the nutrient medium (Fig. 1.19). When the proportion of IAA to kinetin was relatively high, localized regions of

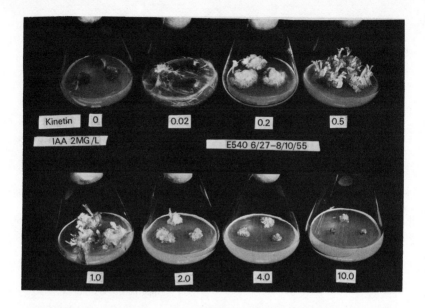

*Figure 1.19.* Regeneration of organs in callus derived from tobacco-pith parenchyma. All cultures illustrated were 44 days old, and were incubated on agar containing necessary nutrients plus 2 mg/l of the auxin, indole-3-acetic acid. The effect of adding a cytokinin, kinetin, at concentrations from 0.02–10 mg/l are shown. With a low cytokinin concentration, root tissues developed, whereas an increase in kinetin concentration to 0.5 mg/l resulted in shoot-bud initiation.

(From *Symposia Soc. Exp. Biol.* **11**, 1957. Original print supplied by Professor F. Skoog.)

a callus differentiated into root apices which grew out as normal roots into the supporting agar medium. A higher concentration of kinetin relative to IAA caused cells to differentiate into shoot apices which developed first into buds and later shoots. Thus, growth and differentiation in callus cells require the presence of both auxin and substances such as kinetin, and the relative amounts of the two can determine the pattern of differentiation in the callus. These discoveries largely disproved older ideas that there were specific hormones required for the initiation of roots, leaves and stems. It is now thought that the control of growth and differentiation in plants involves the interplay of a number of known growth hormones.

Kinetin does not occur naturally in plants, but a number of compounds have been extracted from many plants which have the same sorts of effects in plants as those produced by kinetin. Kinetin and all other compounds which elicit similar physiological responses in plants, or in cultures of plant tissues, are known collectively as *cytokinins*. A cytokinin may therefore be either a naturally occurring hormone in plants, or a synthetic hormone such as kinetin or benzyl adenine (Fig. 1.18).

The chemistry of naturally occurring cytokinins has been intensively studied since about 1960, and most evidence shows that they are adenine (6-aminopurine) derivatives. The first identified natural cytokinin was isolated in the USA from sweet corn (*Zea mays*) kernels by Miller, but its chemical characterization was made by Letham in 1964, working in New Zealand. It was found to be 6-(4-hydroxy-3-methylbut-2-enyl) aminopurine, which for convenience was called *zeatin* (Fig. 1.20). A number of other natural cytokinins have since been isolated from plants and identified: for example, 6-(γ,γ,-dimethylallylamino)purine, which is very similar to zeatin (Fig. 1.20).

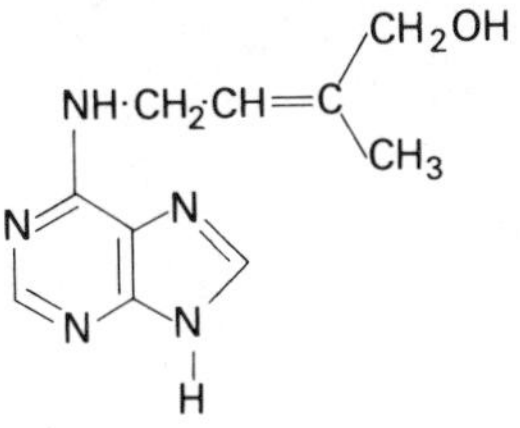

Zeatin
(6-(4-hydroxy-3-methylbut-2-enyl) amino purine)

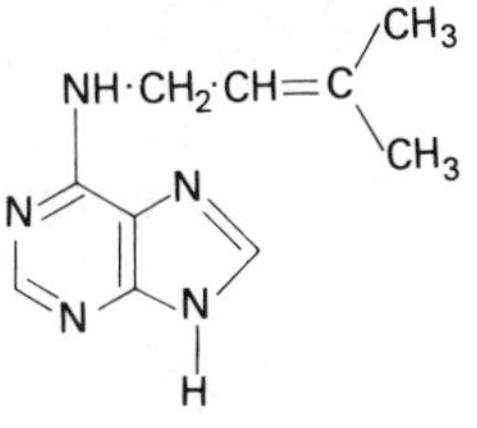

6 -(γ,γ-dimethylallylamino) purine
(also known as, isopentenyl adenine, IPA)

*Figure 1.20*. Structures of two naturally occurring cytokinins of higher plants.

The naturally occurring cytokinins perhaps do not exist in the 'free' form within plant cells, for there is evidence that they are normally bound to a pentose sugar (the riboside) and sometimes the riboside is connected to inorganic phosphate (the ribonucleotide). This means that cytokinins occur in plants as nucleosides or nucleotides. For example, sweet-corn kernels contain zeatin in at least three forms; zeatin, zeatin riboside (a nucleoside) and 9,β-ribosylzeatin-5'-phosphate (a nucleotide) (Fig. 1.21), and there is little doubt that a number of other zeatin derivatives exist in plant tissues.

Nucleoside and nucleotide forms of cytokinins appear to be components of transfer ribonucleic acid (tRNA) in both plant and animal cells, and the possible significance of this in relation to their mechanism of action as growth hormones is the subject of very active research at the present time (p. 161).

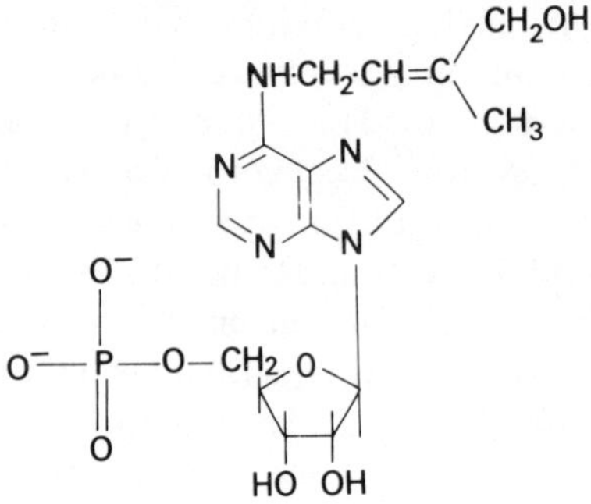

Ribosyl zeatin phosphate
(a nucleotide)

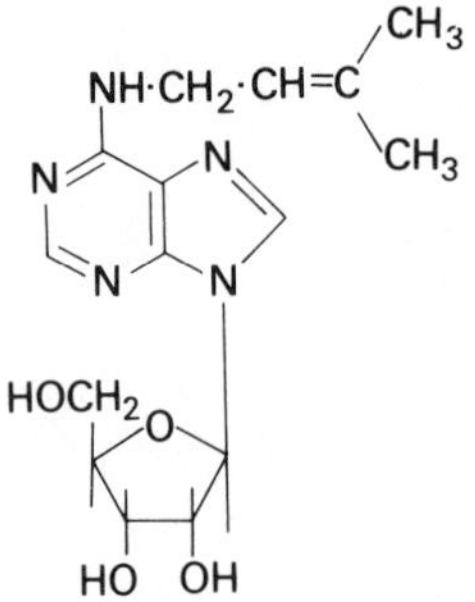

Isopentenyl adenosine (IPA)
(nucleoside form of 6-($\gamma,\gamma$-dimethylallylamino) purine)

*Figure 1.21.* Naturally occurring cytokinins exist in union with a pentose sugar (to form a nucleoside), or with a sugar and inorganic phosphate (forming a nucleotide). This may have significance with respect to their mode of action in cells (see chapter 6). Examples shown are the nucleotide of zeatin (above) and nucleoside of 6-($\gamma,\gamma$-dimethylallylamino) purine.

## Cytokinin metabolism

Little work has yet been done to determine the pathways of natural cytokinin biosynthesis and inactivation. It may, however, be assumed that the earlier stages of their synthesis in plants are the same as those of all purines.

**Cytokinin translocation**

Insufficient attention has yet been devoted to establishing the sites of cytokinin synthesis in plants to allow a simple description of these. It does appear, nevertheless, that the root system is a major region of cytokinin synthesis. This is suggested by a number of observations. Thus, roots are required for the maintenance of protein and chlorophyll levels in leaves, and this dependence on roots can be abolished by providing the leaves with a cytokinin (see also p. 121). Similarly, analysis of the xylem sap ascending from the roots has revealed the presence of considerable quantities of cytokinins. It is thought that the origin of these cytokinins is the root apical meristem. It has not, however, been established whether the shoot system is completely dependent on the roots for a supply of cytokinins, or whether cytokinin synthesis also goes on in some aerial regions of the plant.

Studies of cytokinin translocation have similarly not yet been sufficient to provide us with a clear picture. The evidence referred to in the previous paragraph suggests that cytokinins move upwards, perhaps in the xylem stream, but there are some reports that cytokinin movement in isolated stem or petiole segments is predominantly basipetal.

## Growth inhibitors

Investigations of dormancy phenomena in plants (chapter 5) have revealed the existence of one or more plant hormones which inhibit growth, in contrast to the stimulatory effects of auxins, cytokinins and gibberellins. Because the word 'hormone' literally means 'arousing to activity', some physiologists have objected to classifying a growth inhibitor as a hormone. However, at least one naturally occurring plant growth inhibitor, *abscisic acid* (ABA) (Fig. 1.23) affects plant growth at very low *hormonal concentrations* (of the order of one part per million or less) (Fig. 1.22), and also appears to act as a *chemical messenger* in the regulation of growth. Thus, most scientists now accept that inhibitory growth regulators are also hormones, and interact with the growth promoting hormones such as the auxins, gibberellins and cytokinins.

Soon after auxin was discovered it was suggested that this class of growth hormone may control dormancy. However, no consistent correlation has been found between auxin concentrations in plants and the level of dormancy. There is indeed little to suggest that auxins are directly involved in the control of dormancy. A number of diverse types of chemical (e.g., ethylene chlorhydrin and thiourea) will break the dormancy of many buds and seeds, but applications of auxins either have no effect on the duration of dormancy or they may, in fact, prolong it. Consequently, the idea that dormancy is a result of deficiency in auxin is not tenable. In 1949, Hemberg suggested that dormancy may be attributed to the presence of substances inhibitory to growth. He found that extracts of dormant terminal buds of ash

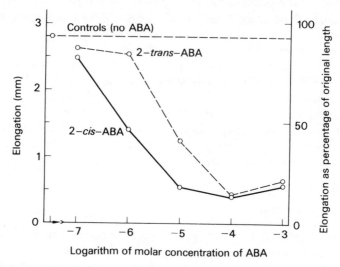

*Figure 1.22.* Inhibition of elongation growth in wheat coleoptile sections (originally 3 mm long) by *cis*- and *trans*-abscisic acid (ABA). $10^{-6}$ M ABA = 0.264 mg/l.
(From J. P. Nitsch, *Ann. N. Y. Acad. Sci.* **144** Art I, ed. Jerome F. Fredrick, 279–294, 1967.)

trees (*Fraxinus excelsior*) contained substances which strongly inhibited the elongation growth of oat (*Avena*) coleoptile segments, and that these growth inhibitors decreased during the winter to reach a very low level by the time dormancy had disappeared. Similar findings were reported by other workers during the early 1950's, which in general supported the concept of endogenous growth inhibitors as controlling agents in dormancy (Figs. 5.1 and 5.2).

Further studies showed that almost all the growth inhibitory activity present in extracts of dormant plant tissues resided in a single highly active substance. The chemical identification of this substance presented a number of problems. Most effort was devoted to the identification of the growth inhibitor present in sycamore (*Acer pseudoplatanus*) leaves and buds, which was given the name *dormin* by Wareing. In 1965, it was established by the British chemist, Cornforth, and co-workers, that dormin was identical to the substance 3-methyl-5-(1'-hydroxy-4'-oxo-2',6',6'-trimethyl-2'-cyclohexen-1'-yl)-*cis*,*trans*-2,4-pentadienoic acid, which had been obtained

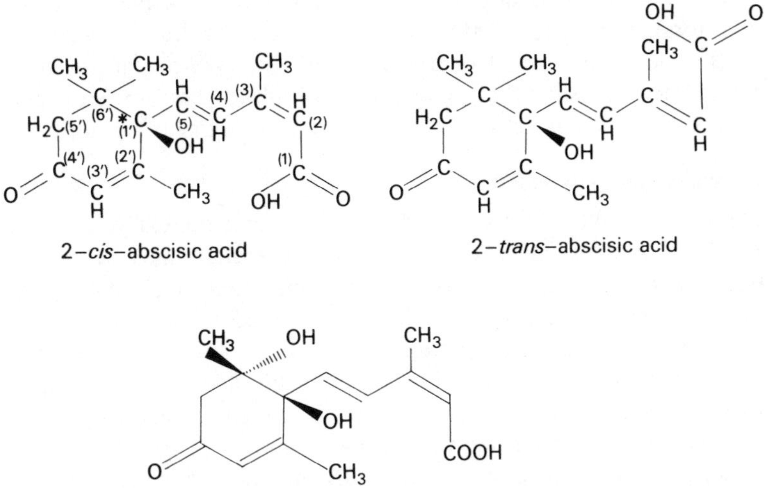

2–*cis*–abscisic acid

2–*trans*–abscisic acid

*Figure 1.23. Top.* Geometric isomers of abscisic acid. Two different configurations (*cis*- and *trans*- forms) are possible around the double bond of carbon atom No. 2 of the side chain. In addition, the carbon atom at the 1' position (marked with an asterisk) is asymmetric (i.e., has four different substituents), which means that abscisic acid exists as either the D(+) or the L(−) enantiomer. *Below.* A (+)-hydroxymethyl breakdown product of ABA formed in plant tissues fed with racemic (±) ABA.

from young cotton fruits earlier the same year by Ohkuma, Addicott and associates in California. Addicott had found that leaf abscission in cotton plants was accelerated by this substance, and hence named it *Abscisin-II*. Two names existed, therefore, for the same compound, but in 1967 it was agreed to call it *abscisic acid*, or ABA in the abbreviated form.

ABA has subsequently been found to be present in many genera and species of plants other than sycamore and cotton. It is a sesquiterpenoid (see p. 28) with the structure shown in Fig. 1.23, and shows both optical and geometric isomerism. Naturally occurring ABA is always in the (+) form, but both the natural (+) and unnatural (−) enantiomers are inhibitory to plant growth. Some evidence suggests that *trans*-ABA is not active (see p. 142). The role of ABA in controlling dormancy has been established (p. 108), and there is already evidence that it is involved in several other developmental phenomena in plants, such as senescence (p. 122), abscission (p. 126), and flower initiation (p. 98).

There seems little doubt therefore that ABA is a plant growth hormone, but a great deal of work remains to be done to establish its functions and how it interacts with the growth-promoting hormones. It is possible that other growth-inhibitor hormones exist in plants, but this has yet to be shown with any degree of certainty.

### Abscisic acid metabolism

*Biosynthesis.* ABA is a sesquiterpenoid, so that the early stages of ABA and gibberellin biosynthesis from mevelonate may be similar (see Fig. 1.15). It has recently been demonstrated that some higher plant tissues can synthesize ABA from added mevalonate. There may be some physiological significance in a common partial pathway of synthesis for ABA and gibberellins, but if so it is not obvious to us at present. In addition, a great number of other compounds are also synthesized along the pathways shown in Fig. 1.15.

Some workers have found that ABA may be formed as a breakdown product of the photo-oxidation of xanthrophylls such as violaxanthin, one of the principal carotenoids in plants (Fig. 1.24). *Catabolism.* (+)-ABA is metabolized in plant tissues to yield at least two metabolites. Work by Milborrow has shown that addition of racemic (±)-[2-$^{14}$C] abscisic acid to tomato shoots results in the rapid appearance of [$^{14}$C]abscisyl-$\beta$-D-glucopyranoside, a (+)-hydroxymethyl compound (Fig. 1.23) and an excess of (−)-[2-$^{14}$C]ABA. It is not known whether this pattern of breakdown of exogenous racemic ABA holds for endogenous (+)-ABA. However, the apparent disappearance of ABA from buds and seeds during their emergence from dormancy (chapter 5) clearly suggests some sort of inactivation mechanism.

### Translocation of abscisic acid

Few direct observations have yet been made of the movement of ABA in plants. Evidence referred to later (p. 108) that leaves of woody plants exposed to short days exert an inhibitory effect on the growth of the shoot apical region, indicates that ABA is manufactured in mature leaves and can move up the stem to the shoot apex. By analogy with what is already known of the movement of solutes from mature leaves, we might expect that transfer of ABA to the shoot apex takes place in the phloem and perhaps xylem as well. Analysis of the contents of phloem sieve tubes (obtained by collected 'honeydew' from aphids feeding on willow stems) showed ABA to be present. Similarly, the xylem sap was also found to contain ABA. Thus, what slight evidence is to hand suggests that translocation of ABA goes on in the phloem and xylem along with other organic and inorganic solutes. We have no knowledge of the velocity of ABA translocation.

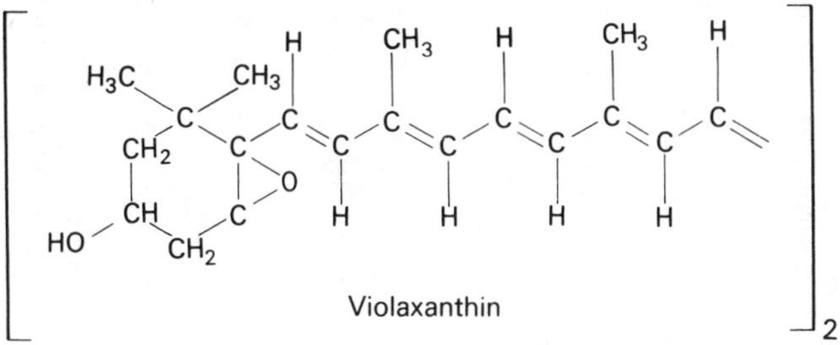

Violaxanthin

Figure 1.24. Structure of violaxanthin. This naturally occurring carotenoid, and perhaps also other carotenoids, may serve as a precursor for the biosynthesis of abscisic acid. Synthesis of abscisic acid from such carotenoids could involve oxidative cleavage of each $C_{40}$ carotenoid molecule to yield eventually two fragments, each of fifteen carbon atoms (i.e., to sesquiterpene molecules, each with the structure of abscisic acid).

## Ethylene

Ethylene (Fig. 1.25) is a gas at all temperatures under which plants can survive. If we consider that ethylene is a plant growth hormone, then we have a somewhat surprising situation in that it is a gaseous hormone. When one remembers that other hormones serve as correlation factors (p. 3), it becomes difficult to visualize how a gas can be distributed in a plant with a sufficient degree of precision to allow it to function in the integration of growth in different parts of the organism. Nevertheless, considerable evidence has now been adduced which supports the view that ethylene is a plant hormone.

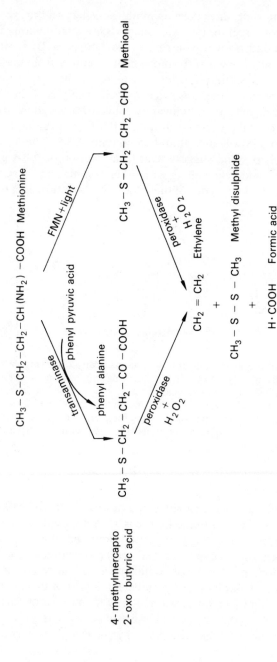

Figure 1.25. Ethylene, and two alternative possible pathways of its biosynthesis in plant tissues from the amino acid, methionine. Other possible precursors of ethylene include β-alanine.

The problem of how its movement is controlled within the plant body remains, but might be less serious than imagined. After all, although auxin movement follows fairly strict patterns (p. 17), the translocation of gibberellins shows no such restraint (p. 28), and yet we accept that both may serve as chemical messengers. In addition, at the low concentrations in which ethylene occurs in plants, most of the substance will be dissolved in the aqueous phase of cells and cell walls.

The first step in assigning the label 'growth hormone' to any substance, is to establish that it possesses the property of influencing growth whilst at the same time not serving as a nutrient. The concentrations of ethylene which have been found to influence physiological processes are so low (down to $0.06 \, \mu g/l$) as to discount the possibility that it serves a nutrient.

It has been known for many years that ethylene is evolved by plants. Fairly large amounts of the gas are given off during the ripening of fruits, and the maximum rate of ethylene production occurs during the period of maximum respiration rate (i.e., during the *respiratory climacteric* which takes place just before senescence in many fruits, Fig. 5.11). It appears, in fact, that ethylene production is intimately associated with the ripening process in fruits, for many fruits which do not normally exhibit a respiratory climacteric do so when exposed to ethylene, and their ripening rates are greatly increased (Fig. 5.17). This effect of ethylene is utilized in the citrus industry, where oranges, lemons and grapefruits are sometimes picked whilst still green and ripened in gas chambers containing ethylene. The ethylene gas induces accelerated conversion of starch to sugars and hydrolysis of pectin within the flesh of the fruits, and also speeds up the associated color changes at the fruit surface.

Ethylene is now known to be produced in plant organs other than fruits. In the case of dark-grown etiolated pea seedlings, it has been observed that growth in the upper plumular region is inhibited, and that this is a result of the upward diffusion of ethylene from its site of synthesis in a region of the epicotyl just below the plumular hook. On exposure to light, ethylene production falls and at the same time growth of the plumular region picks up. This finding provides strong evidence that endogenous ethylene acts as a correlative growth inhibitor. However, many experiments have demonstrated effects of exogenous ethylene on plants and, in general, responses to ethylene are similar to those which are induced by *high* auxin concentrations.

A number of experiments have shown that ethylene production in plant tissues rises sharply when the auxin concentration in the same tissues is high. The actual concentrations of auxin which elicit this effect vary, but are always at such a level that they are supra-optimal and inhibitory to growth (see p. 11).

We will consider where ethylene 'fits in' to the picture of hormonal control of plant growth and development in later chapters, along with similar considerations of the other growth hormones.

**Ethylene metabolism**

The amino acid, methionine, appears to be the immediate precursor of ethylene in plants (Fig. 1.25), but very little else is known of the biochemistry of ethylene synthesis or inactivation.

## Defining plant growth hormones

We may define a growth hormone as a substance which is produced in one region of a multicellular organism, and is then transferred in very small quantities to other parts where growth responses occur. As a general description of animal and plant hormones this is satisfactory. In the case of plant growth hormones, however, it is virtually impossible to give a succinct definition of each class of growth hormone. This is due to the fact that specific physiological effects are not reserved to any one particular class of hormone, and also that, in the case of auxins at least, a bewildering variety of chemicals possess auxin-like biological activity. However, lack of precise definitions does not prevent us from knowing what we mean when we say 'auxin' or 'gibberellin', etc. To a large extent our concept of each class of plant growth hormone is based upon a combination of the chemical constitution *and* physiological effects of its members. It would be very much easier, and more satisfactory, to use a system of nomenclature based upon both the functions of plant growth hormones and their site of action in individual cells. However, until such day arrives that we possess the necessary information, we are compelled to adopt the existing imprecise, and often vague, definitions.

**Auxins**

These are all weak organic acids, with the acidic group situated at the end of a side-chain attached to an unsaturated ring or ring system (e.g., IAA has an acetic acid side-chain attached to an indole ring system, Fig. 1.3). Non-acidic substances (e.g., IAN, Fig. 1.3) which exhibit auxin activity probably do so by virtue of enzymic conversion to acidic substances. Auxins either enhance stem and coleoptile extension growth or inhibit it, depending on the concentration applied. The effect of exogenous auxin on growth can be seen only if the bioassay material is separated from the natural source of auxin (e.g., by cutting off a coleoptile tip and measuring the effect of applied auxin on growth of the lower tissues). As will be seen later, auxins have many effects in addition to those on extension growth.

### Gibberellins

This class of hormones also consists of a number of compounds each of which is acidic. All are based upon the *gibbane carbon skeleton* (Fig. 1.13), which is closely related to diterpenes. A typical effect of gibberellins is that their application to *intact* plants often results in increased stem and leaf extension growth. This effect of gibberellins contrasts strongly with the auxins, which rarely enhance growth when sprayed on an intact plant. On the other hand, isolated coleoptile or stem segments elongate little in response to applied gibberellins, whereas, as we have seen, they do respond to exogenous auxins. In addition to effects on growth, gibberellins influence various aspects of plant development, such as dormancy, senescence and flowering.

### Cytokinins

These were discovered as substances essential for the maintenance of cell division activity in sterile cultures of plant tissues. They also interact with auxins in determining the pattern of differentiation in a callus. In addition to effects in callus cultures, we are now aware that cytokinins are involved in many other physiological processes in whole plants, such as apical dominance and senescence. In contrast to the other classes of growth-promoting hormones (auxins and gibberellins), cytokinins are chemically basic, rather than acidic, in nature. Endogenous cytokinins all appear to be derivatives of the purine nitrogenous base, adenine.

### Growth inhibitors

Only one naturally occurring plant hormone which inhibits rather than promotes growth has so far been identified. This is abscisic acid, some of the effects of which have been described earlier (p. 38). It is therefore premature to devise any general definition of growth

**43**

inhibitors. If, however, other growth-inhibitory hormones operate in plants, it is likely that they, as abscisic acid appears to do, will interact with growth-promoting hormones and have effects on plant development in addition to simply inhibiting growth.

## Ethylene

We have no information which suggests that any gases other than ethylene function as hormones in plants. Ethylene therefore stands on its own, and no useful purpose is to be served by attempting to define it in terms of its physiological functions or chemical relationships with other compounds.

## Further reading list

1. Cleland, R. E. 'The Gibberellins', in *The Physiology of Plant Growth and Development* (ed. M. B. Wilkins), pp. 49–81, McGraw-Hill, London, 1969.
2. Fox, J. E. 'The Cytokinins', in *The Physiology of Plant Growth and Development* (ed. M. B. Wilkins), pp. 85–123, McGraw-Hill, London, 1969.
3. Goldsmith, Mary Helen M. 'Transport of Plant Growth Regulators', in *The Physiology of Plant Growth and Development* (ed. M. B. Wilkins), pp. 127–162, McGraw-Hill, London, 1969.
4. Heslop-Harrison, J. 'Plant Growth Substances', in *Vistas in Botany* Vol. 3. *Recent Researches in Plant Physiology* (ed. W. B. Turrill), pp. 104–194, Pergamon Press, London, 1963.
5. Lang, Anton. 'Gibberellins: Structure and Metabolism', *Ann. Rev. Plant Physiol.* **21**, 537–570, 1970.
6. Leopold, A. C. *Plant Growth and Development*, McGraw-Hill, N.Y., 1964.
7. Mapson, L. W. 'Biosynthesis of Ethylene and the Ripening of Fruit', *Endeavour* XXIX, 29–33, 1970.
8. Paleg, L. G. 'Physiological Effects of Gibberellins', *Ann. Rev. Plant Physiol.* **16**, 291–322, 1965.
9. Skoog, F., and D. J. Armstrong. 'Cytokinins', *Ann. Rev. Plant Physiol.* **21**, 359–384, 1970.
10. Shantz, E. M. 'The Chemistry of Naturally-Occurring Growth-Regulating Substances', *Ann. Rev. Plant Physiol.* **17**, 409–438, 1966.
11. Steward, F. C. *Growth and Organization in Plants*, Addison-Wesley Publ. Co., Reading, Mass., 1968.
12. Thimann, K. V. 'Plant Growth Substances; Past, Present, and Future', *Ann. Rev. Plant Physiol.* **14**, 1–18, 1963.
13. Wareing, P. F. and I. D. J. Phillips. Chapters 1–4 in *The Control of Growth and Differentiation in Plants*, Pergamon Press, Oxford, 1970.
14. Wareing, P. F., and G. Ryback. 'Abscisic Acid: A Newly Discovered Growth Regulating Substance in Plants', *Endeavour* XXIX, 84–88, 1970.

# 2. Growth hormones in shoot and root development

## Shoot growth

A typical vegetative shoot consists of an apical meristem, a stem, and leaves. In dicotyledonous plants, lateral buds are also present which may grow out to form lateral branches (see p. 60). All these structures arise from the meristematic cells of the apex, and it is clearly of great interest to students of plant biology to understand how the diverse tissues and organs of the shoot differentiate in an ordered manner from originally identical cells at the stem tip. It must, however, be admitted from the outset that there are enormous gaps in current knowledge of the controls operating at the shoot apex itself. Some cells formed in the apical meristem remain permanently meristematic and are retained in the apical meristem, others develop into the procambial strands and also retain meristematic activity. Others, again, lose their ability to divide either immediately after being cut off from the apical meristem, or soon afterwards, and develop into stem parenchyma, leaf mesophyll cells, epidermal cells, etc. It is possible that the destiny of cells formed at the apex is determined by a pattern of varying concentrations of nutrients and growth hormones, but there is no good evidence to support this proposition, for it is exceedingly difficult to put to experimental test owing to the very small size of the structures under consideration. We cannot, therefore, ascribe any role for growth hormones in the shoot apical meristem itself, although the possibility cannot be discounted either.

Immediately below the apical meristem proper (sometimes called the *eumeristem*), in almost all plants, is a region of continued cell division activity which we may refer to as the *sub-apical meristem* (Fig. 2.2). There is a preponderence of transverse cell divisions in the sub-apical meristem, giving rise to the parallel files of cells typical of pith and cortical tissues. Mitotic activity in the sub-apical meristem continues for at least one to two centimeters, and sometimes much further, below the eumeristem (Fig. 2.2). In many monocotyledonous plants, particularly in grasses and cereals, the sub-apical meristem extends the whole length of the stem. It may be broken up into separated 'islands' of meristematic tissue during stem development, which leads to the appearance of the classical *intercalary meristems*.

**Stem elongation**

The shoot of a plant elongates by two processes: (i) addition of new cells to the stem by mitotic activity in the eumeristem and sub-apical meristem, and (ii) enlargement of these cells, principally as a result of vacuolation. The vacuolation phase of stem elongation growth is characterized by the individual cells enlarging much more along the axis of the stem than transversely. For this reason we often speak of *cell-extension growth* in an elongating organ such as a stem.

Any consideration of the role of growth hormones in the control of stem elongation growth must therefore take into consideration both cell division and enlargement processes. It has been stated above that we have no good evidence for growth hormones operating in the eumeristem. By contrast, there is reason to believe that at least part of the effect of exogenous gibberellins in stimulating stem elongation (p. 25) is due to an enhancement of cell-division activity in the sub-apical meristem. Microscopic studies of stems of various species of plants have shown that the sub-apical meristem is more important than the eumeristem in providing new cells which contribute to stem elongation. For this reason the sub-apical meristem has been termed a 'primary elongating meristem' by R. M. Sachs.

Many plants grow taller when sprayed with a gibberellin solution. The greatest growth response to applied gibberellins occurs in genetically dwarfed plants (Fig. 2.1) such as certain varieties of pea, bean and corn (*Zea mays*) (i.e., mutant varieties which normally remain short when compared with related tall varieties, because they possess a 'dwarfing' gene or genes). The question arises as to whether gibberellins induce dwarf plants to become phenotypically tall by

*Figure 2.1.* Effect of gibberellic acid (GA$_3$) on normal tall, and on dwarf mutant d$_5$ *Zea mays* plants. From the left: untreated tall; GA$_3$-treated tall; untreated dwarf; GA$_3$-treated dwarf. Plants treated with GA$_3$ received a total dose of 250 micrograms.
(From B. O. Phinney and C. A. West, 'Gibberellins and the Growth of Flowering Plants', in *Developing Cell Systems and Their Control* (ed. D. Rudnick), Ronald Press Co., 71–92, 1960. Original print provided by Dr. B. O. Phinney.)

influencing cell division or cell enlargement in the stem. Stems of genetic dwarf plants contain both fewer and shorter cells than tall varieties. Treatment with a gibberellin causes cell number and cell length to increase to the normal (tall) values. The increase in cell number which takes place in the response of dwarf plants to gibberellins results from stimulated cell-division activity in the sub-apical meristem. Similar observations have been made on other species, particularly 'rosette' plants (i.e., those whose leaves are arranged in a rosette around an unelongated stem, but which exhibit 'bolting' of the previously compressed stem just prior to flower formation) such as *Hyoscyamus niger* and *Samolus parviflorus*. Here, exogenous gibberellins cause the appearance of a 'new' sub-apical meristem below the eumeristem (Fig. 2.2). Normally, sub-apical meristematic activity in rosette plants takes place only in response to exposure to long photoperiods, when flowering occurs. It is now known that gibberellins activate sub-apical meristematic activity in tall caulescent plants, as well as in rosette and genetic dwarf plants (Fig. 2.3).

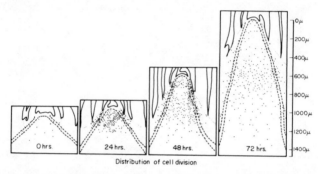

Distribution of cell division

*Figure 2.2.* Stem elongation and sub-apical meristematic activity in a rosette plant following application of gibberellic acid at 0 hr. Semi-diagrammatic sketches of longitudinal sections through the stem axis of *Samolus parviflorus*. Each dot represents one cell undergoing mitosis seen in a 64 μm thick slice.
(From R. M. Sachs, C. Bretz, and A. Lang, *Exper. Cell Res.* **18**, 230–244, 1959.)

There is, therefore, good reason to believe that stem elongation is at least partially controlled by the rate of cell division activity in the sub-apical meristem, and that this, in turn, is determined by the quantity of gibberellins available. There is, however, little to suggest that other growth-promoting hormones, such as auxins or cytokinins, are active in the same way as the gibberellins in the sub-apical meristem, although their presence may modify the response of this region to gibberellins.

What, now, can we say of the other component of stem elongation, cell-extension growth? There is no doubt that although the number of cells present in a stem is very important in determining stem length, overall stem elongation is largely a reflection of cell extension growth. This is because vacuolating cells enlarge to many times their original size (e.g., in some bamboo stems, newly formed cells at the apex are 8 to 10 μm long, but they extend during vacuolation to about 100 μm). One may therefore regard the eumeristem and sub-apical meristem as providing capital, in the form of new cells, on which a very high rate of 'interest' is earned by cell enlargement. The

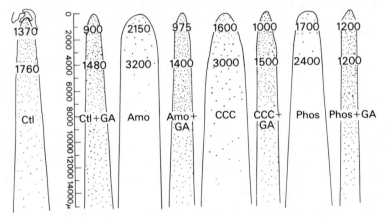

*Figure 2.3.* Density and distribution of sub-apical meristematic activity in stems of *Chrysanthemum morifolium* following treatment with gibberellic acid (GA), growth retardants (Amo, CCC, and Phosfon-D), and combinations of GA with retardants. Control plants untreated ('Ctl' at extreme left). Each dot represents one mitotic figure in a 60 μm thick longitudinal slice. The numbers at 1 mm and 4 mm below the apex are the transverse diameters of the pith tissue in μm at these levels. Mitotic activity was greatly reduced by the growth retardants but transverse growth increased. GA increased mitotic activity, primarily 6 mm and more below the apex, and reduced transverse growth. The effects of the growth retardants and GA were antagonistic.
(From R. M. Sachs and A. M. Kofranek, *Amer. J. Bot.* **50**, 772–779, 1963.)

control of cell-extension growth in stems is known to be achieved through the agency of both auxins and gibberellins. Much of the evidence implicating these growth hormones in extension growth has come from studies of coleoptiles (which are modified, tubular leaves), but, in general, the findings seem to be applicable to stems as well.

Etiolated coleoptiles of oat or wheat are particularly useful experimental subjects for studying cell-extension growth, for by the time they attain a length of about 2 cm, cell-division activity ceases. Coleoptile elongation after this time depends solely upon elongation of existing cells. We have already seen (p. 7) that the coleoptile tip synthesizes auxin which is required for elongation of the whole organ. The necessity of apically synthesized auxin for cell-extension growth has been confirmed by the results of many experiments using segments of coleoptiles cut out from seedlings. Such segments are deprived of their normal source of auxin by removal of the coleoptile tip. Excised etiolated coleoptile segments bathed in water, or a nutrient solution, elongate only slightly, but when surrounded by a solution containing an auxin such as IAA at concentrations of from 1 to 10 mg per liter (approximately $5 \times 10^{-6}$ M to $5 \times 10^{-5}$ M IAA) their length may double or treble within twelve hours (Fig. 1.4). This demonstrates very clearly the promotive effect of sub-optimal or optimal concentrations of auxins upon cell enlargement. Higher concentrations of auxins (e.g., $10^{-4}$ molar or above) usually inhibit elongation, and are therefore supra-optimal (Fig. 1.4). Coleoptile segments similarly treated with gibberellins, on the other hand, show either no or a more limited response. This does not, however, mean that gibberellins have no influence upon cell-extension growth, for when a gibberellin is supplied to coleoptile segments along with an auxin then the elongation which takes place is often significantly greater than when only auxin is supplied. It consequently appears that both auxin *and* gibberellin are required for maximum cell-extension growth, and perhaps that gibberellins are unable to induce cell extension in the absence of an auxin. Auxins and gibberellins therefore interact in the control of cell-extension growth in coleoptiles. It is possible that gibberellins inhibit auxin-induced ethylene biosynthesis (see p. 42). If this is so, then less ethylene (which inhibits cell elongation) would be produced with a given auxin concentration when a gibberellin is also present.

The effects of applied auxins and gibberellins on elongation growth in stem segments (i.e., segments of young, incompletely elongated, internodes taken from just below the apical bud) are similar to those described for coleoptile segments. That is, they elongate in response to auxins and even more to a mixture of auxins and gibberellins but not to gibberellins alone (Fig. 2.4). The inability of isolated internode segments to elongate in the presence of gibberellins might appear strange when one recalls that treatment of *intact* plants with gibberellins result in enhanced internode elongation, and that part of this response is due to increased cell extension (p. 47). The reason for this apparent paradox involves the fact that whereas isolated stem segments have been deprived of their principal source of auxin, internodes of intact plants receive a supply of auxin

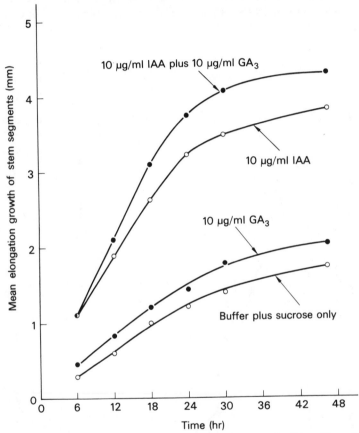

*Figure 2.4*. Effects of indole-3-acetic acid (IAA) and gibberellic acid (GA₃) on elongation growth in excised segments of pea internode. Auxins (IAA) enhance elongation growth of excised segments much more markedly than gibberellins (GA₃). The addition of both auxin and gibberellin results in either an additive or sometimes synergistic effect on growth of segments.
(From P. W. Brian and H. G. Hemming, *Ann. Bot.*, N.S. **22**, 1–17, 1958.)

from the young leaves (or even from cotyledons in very young seedlings). This again suggests that gibberellin-enhanced growth occurs only when auxin is also present.

We may reasonably assume, therefore, that the internal mechanism which controls stem-extension growth involves both auxins and gibberellins. This view is supported by results of experiments which have shown that the levels of endogenous auxins and gibberellins are highest in just those regions of the stem which are elongating most rapidly (Figs. 2.5 and 2.6).

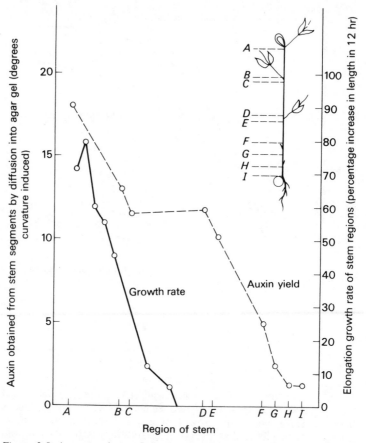

*Figure 2.5.* A comparison of the distribution of elongation growth and diffusible auxin in the pea epicotyl.

(From T. K. Scott and W. R. Briggs, *Amer. J. Bot.* **47**, 492–499, 1960.)

Cytokinins have not been shown to play any part in the control of stem elongation, although the importance of this class of hormone in regulating cell division in callus cultures (p. 30) leads one to suspect that endogenous cytokinins may be involved in determining cell-division activity in the eumeristem and sub-apical meristem. In the absence of good experimental evidence, however, this must remain for the moment no more than a suspicion.

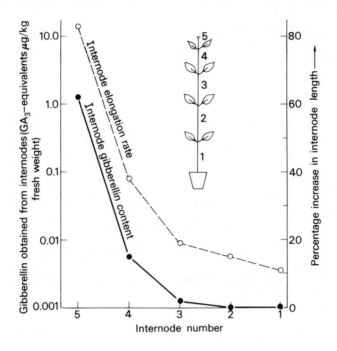

*Figure 2.6*. A comparison of the distribution of elongation growth and diffusible gibberellin in the sunflower (*Helianthus annuus*) stem. (From R. L. Jones and I. D. J. Phillips, *Plant Physiol.* **41**, 1381–1386, 1966.)

Over the past few years, a number of investigators have been led to conclude that ethylene as well as auxins and gibberellins is involved in the hormonal control of stem elongation. Ethylene inhibits internode elongation (Fig. 2.7), and in the case of pea seedlings there is good evidence that it can serve as a correlative elongation-growth inhibitor (see chapter 1, p. 41). It now seems likely that growth

**53**

inhibition by supra-optimal concentrations of auxin is the result of enhanced ethylene biosynthesis (see p. 42 and Fig. 1.4).

Natural plant-growth inhibitors such as abscisic acid (ABA) inhibit elongation of excised coleoptile or stem segments (Fig. 1.22), and also internode elongation when applied to intact plants (p. 108).

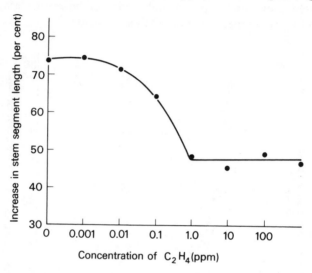

*Figure 2.7.* Inhibition of elongation growth in excised pea stem segments by ethylene.
(From S. P. Burg, *Regulateurs Naturels de la Croissance Végétale*, Édition de la Rech. Sci. 1964, 718–724.)

In general, it has been found that ABA induces developmental responses in plants opposite to those elicited by the gibberellins, and sometimes by cytokinins. It is possible, therefore, that endogenous ABA is involved along with gibberellins and cytokinins in the regulation of sub-apical meristemic activity. This suggestion is supported by the inhibitory influence that certain synthetic *growth retardants* such as CCC, AMO and 'Phosfon-D' (p. 28) have upon cell-division rates in the sub-apical meristem. The effects that these growth retardants have on plants are in some ways similar to those produced by naturally occurring ABA, and they not only reduce normal sub-apical meristematic activity, but they also antagonize the cell-division promoting effect of exogenous gibberellins upon this region of the stem (Fig. 2.3). It is, therefore, possible that endogenous gibberellins and ABA similarly interact to regulate cell division rates in the sub-apical meristem.

## Radial growth and differentiation in the stem

Stems are usually tapered along their length, the basal region having a distinctly greater diameter than more apical parts. Thus, in addition to elongation, transverse, or radial, growth occurs in stems.

Radial stem growth occurs as a result of several processes. In rapidly elongating stems, new cells formed in the sub-apical meristem enlarge mainly in the longitudinal direction, and although transverse expansion of the cells goes on at the same time it contributes only very slightly to stem diameter. Conversely, when elongation growth is slow a greater proportion of cell enlargement takes place in the horizontal plane, transverse to the stem axis. Thus, stems of etiolated or gibberellin-treated plants are longer and thinner than the stems of light-grown or growth-retardant treated plants. It appears that factors which suppress the rate of cell division in the sub-apical meristem also cause increased radial expansion of the stem. These factors include synthetic growth retardants (Fig. 2.3), abscisic acid, *high* auxin concentrations, ethylene, and perhaps cytokinins, as well as environmental factors such as light and low temperatures. We do not yet understand how the direction of cell enlargement is controlled, but since growth hormones can influence the rate of cell division in the sub-apical meristem, and also the rate of cell extension growth, they will clearly play some part in the process.

Increase in diameter of the stem in dicotyledenous species is also achieved through the activities of the *lateral meristems*, of which the most important in transverse stem growth is the *vascular cambium*. New cells produced by the vascular cambium normally differentiate into phloem and xylem vascular tissues. Radial stem growth resulting from vascular cambial activity is most obviously seen in woody shrubs and trees, in which so-called secondary thickening goes on year after year. Successive layers of lignified xylem vessels and tracheids are formed to the inner face of the vascular cambium, and phloem tissues to the outer side. However, the first suggestion that growth hormones serve as controlling agents in vascular cambium activity came from work by Snow in 1935 on sunflower (*Helianthus annuus*), a herbaceous annual. Decapitation of the shoot (i.e., excision of the apical bud) resulted in cessation of cell-division activity in the fasicular cambium of the uppermost young internode, and also prevented the usual formation of an interfasicular cambium. Snow discovered that the vascular cambium requires a supply of auxin, which normally comes from the young leaves of the apical bud. Thus, normal cambial activity and secondary thickening took place in decapitated sunflower plants only when auxin (IAA) was applied to the upper cut end of the stem.

Apically synthesized auxin is also essential for cambial activity in woody species. In temperate zone trees, the cambium is inactive during wintertime, but renewed cell-division activity commences at the base of expanding buds in spring. A wave of cambial activity then gradually spreads downwards through the twigs to the branches, and down the trunk. A supply of auxin from expanding buds is clearly instrumental in the initiation of cambial activity in spring, for disbudding prevents the onset of cell division in the cambium, but application of auxin to the upper end of disbudded twigs results in normal basipetal activation of the cambium. Thus, both auxin movement (p. 17) and the initiation of cambial activity is basipetal in stems.

Gibberellins, and probably cytokinins, in addition to auxins, are also concerned in the control of cell division activity in the vascular cambium. Wareing found that treatment of the upper end of disbudded twigs of sycamore (*Acer pseudoplatanus*) and poplar (*Populus robusta*) with either IAA or $GA_3$ promoted cambial activity, but that most active cell division occurred when *both* the auxin and gibberellin were supplied. Further, completely normal differentiation of xylem and phloem took place only when auxin *and* gibberellin were present (Fig. 2.8). When only IAA was applied to the twigs, very little phloem differentiated and only a few of the new cells formed to the inside of the cambium differentiated into lignified xylem vessels. With $GA_3$, more phloem was formed, but although a wide band of new cells was formed to the inner face of the cambium, none of these differentiated into xylem vessels. Application of both IAA and $GA_3$ resulted in much more normal differentiation of both phloem and xylem tissues.

Thus, auxin and gibberellin control both the rate of cell division in the vascular cambium, and the pattern of differentiation of newly formed cells. We may reasonably assume that the auxin comes from the young leaves of the shoot, but we can be less certain of the origin of endogenous gibberellins concerned in the control of the cambium region. Studies on herbaceous plants have shown that gibberellins,

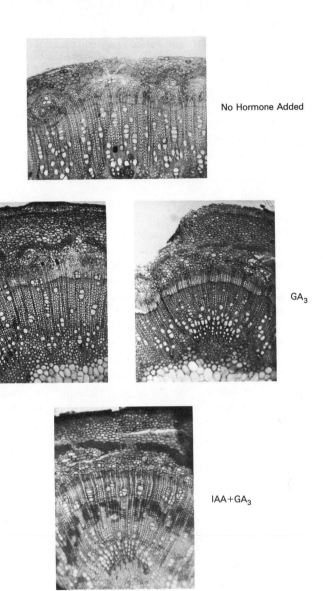

No Hormone Added

IAA

GA₃

IAA+GA₃

*Figure 2.8*. Effect of auxin (IAA) and gibberellin (GA₃) on cambial
activity and xylem differentiation in nondormant disbudded twigs of
poplar (*Populus robusta*).
(Original prints supplied by Professor P. F. Wareing, F.R.S.)

as well as auxins, are synthesized in young leaves (Fig. 1.14), so that it is possible that these, too, are supplied to the cambium by expanding buds in trees.

A role for cytokinins in the control of cambial activity is suggested by the finding that treatment of excised pea-stem segments with cytokinin resulted in a stimulated rate of division in the cambium, and increased lignification of differentiating xylem elements. More work is needed, however, before any conclusions are drawn.

### Leaf growth

It is impossible to give any clear indication of the roles of growth hormones in the growth of leaves. Because growth in developing leaves involves cell division, cell enlargement and differentiation, it is likely from what we know of the effects of growth hormones on these processes in other organs that they are also functional in growing leaves. Very little definite evidence is available, however, to substantiate this belief.

Application of growth hormones to leaves has shown that, in general, auxins inhibit mesophyll expansion but promote elongation of veins, whereas gibberellins and cytokinins can stimulate both vein and mesophyll growth. The auxin and gibberellin contents of leaves are positively correlated with their growth rates, which provides good circumstantial evidence that these hormones are somehow involved in the control of leaf growth. There is no doubt that leaves are very active centers of auxin and gibberellin synthesis for, as we have seen, stem tissues are dependent upon leaves for a supply of these growth hormones. Also, the quantities of auxins and gibberellins which are actually exported to the stem by leaves of various ages have been measured for several species. The way in which this can be done is to detach individual leaves from the stem, and stand each with the cut end of petiole against the surface of a block of agar-gel for a few hours. Both auxins and gibberellins, which are synthesized in the lamina, diffuse out from the end of the petiole into the agar, and the amounts so obtained can be determined by bioassay. It is found that very young expanding leaves are the most active sites of auxin and gibberellin synthesis, and that there is a progressive fall in their activity in this respect as they mature (Fig. 2.9).

There is no evidence which suggests that cytokinin synthesis occurs in leaves. In fact, certain evidence points in the opposite

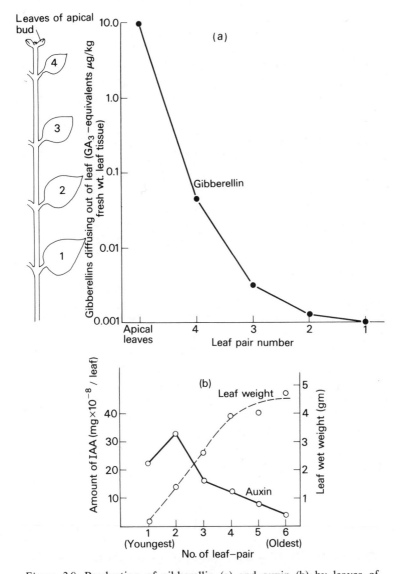

*Figure 2.9.* Production of gibberellin (a) and auxin (b) by leaves of various ages. Young expanding leaves produce, and export to the stem, considerably greater amounts of both gibberellins and auxins than older mature leaves. (a) sunflower leaves. (b) *Coleus* leaves.

(a) from R. L. Jones and I. D. J. Phillips, *Plant Physiol.* **41**, 1381–1386, 1966. (b) from R. H. Wetmore and W. P. Jacobs, *Amer. J. Bot.* **40**, 272–276, 1953.

**59**

direction, for detached leaves often need to be supplied with a cytokinin to remain green and healthy (p. 121).

Whereas gibberellins and auxins are produced most actively in young leaves, the growth inhibitor abscisic acid is synthesized in mature leaves (p. 108), and may be involved in the processes of leaf senescence and abscission (chapter 5).

### Apical dominance in shoots

In most plants the apical bud of the shoot exerts an inhibitory influence on the growth of lower axillary buds. The main stem extends upward more rapidly than side branches grow out, and axillary buds are said to be subject to *correlative inhibition* by the apical bud. The degree of apical dominance in shoots varies between species and varieties, so that in some plants (e.g., sunflower, *Helianthus annuus*) apical dominance is very strong and the axillary buds hardly grow at all, whereas in others axillary buds grow out much more freely and the shoot system therefore assumes a more bushy appearance (e.g., tomato, *Lycopersicum esculentum*). Also, correlative inhibition of axillary buds diminishes as a plant gets older. This is most clearly illustrated in trees, where strong growth of the main stem occurs early in their lives, followed by a gradual reduction in apical dominance over the years which leads eventually to the formation of a rounded 'crown' consisting of a finely branched shoot system with no obvious leading shoot. Additional variability in the strength of apical dominance can be seen when the nutritional status of the plant is modified, for when nutrients such as nitrogen are in plentiful supply, axillary buds are less inhibited by the presence of the apical bud.

Apical dominance, or correlative bud inhibition, provides a good example of the correlative forces which are at work in a growing plant. Plant physiologists have attempted to unravel the mechanisms involved in the phenomenon, for it is hoped that knowledge of the physiological basis of apical dominance will shed light upon the various other examples of growth correlations which can be seen in plants. In some cases growth correlations are characterized by one part of the plant exerting a stimulatory influence upon growth in another part (e.g., the coleoptile tip synthesizes auxin which promotes growth in the lower part of that organ), but in other examples, such as apical dominance, the effect of one region of the plant is to inhibit growth in another part.

Growth hormones are, by definition, correlation factors (i.e., they are chemical messengers), and it is reasonable to consider the possibility that they are involved in the mechanism of apical dominance. However, at the time that scientists first took an interest in apical dominance the existence of plant growth hormones had not been established. Early investigators of this problem were drawn by the paradox that axillary buds are situated nearer the root system and mature photosynthesizing leaves than the apical bud, and yet grow less vigorously than the latter. They noted that removal of the apical bud ('decapitation') resulted in immediate outgrowth of the axillary buds, thus demonstrating that arrested growth of axillary buds lasts only so long as an actively growing apical bud is present. Prior to the discovery of auxin, preferred explanations of apical dominance were based upon the idea that competition for nutrients takes place between actual and potential centers of growth. It was assumed that axillary buds fail to elongate because they are starved of inorganic and organic nutrients. This *nutritive theory* of apical dominance held that the apical bud receives nutrients in greater quantities than do axillary buds. The principle underlying this theory is that nutrients may be expected, in accordance with physical laws, to move in response to their concentration gradients. Thus, active growth activity in an apical bud constitutes a 'sink' for nutrients, toward which these will then move from mature leaves and roots. It was envisaged that once this process is set in motion at germination, then later-formed axillary buds never get a chance to become active metabolic sinks and remain starved.

The discovery of auxin, and the realization that this class of growth hormone is produced in young leaves of apical buds, led Thimann and Skoog in 1934 to perform a simple experiment with bean plants (*Vicia faba*) which revolutionized ideas of the basis of apical dominance. These two plant physiologists discovered that axillary buds failed to grow out even in decapitated plants when IAA was applied to the upper cut end of the stem. In other words, auxin acted as a substitute for the apical bud in correlative inhibition of axillary buds. This observation has been confirmed for many species of plants (Fig. 2.10), and it is now clear that an extremely important part of the apical-dominance mechanism involves the production and transmission of auxin by the young growing leaves of the apical bud.

In contrast to other natural effects of auxin in plants, apical dominance is a phenomenon which involves inhibition of growth by activities of auxin. The means by which axillary buds are held in check by apically synthesized auxin is still a topic of active research. At one time it was thought that axillary buds, like roots (Fig. 2.14), are particularly sensitive to auxin, so that concentrations of auxin which will promote main-stem growth would be supra-optimal and therefore inhibitory to buds. This was known as the *direct theory* of auxin inhibition of axillary buds, but has been superseded in recent years for a number of reasons which we need not go into here, except to mention that auxin levels in correlatively inhibited axillary buds are very low and rise markedly when they are released from inhibition by excision of the apical bud. It is now believed that auxin exerts inhibitory effects on axillary buds in an *indirect* manner.

*Figure 2.10.* Apical dominance in runner bean (*Phaseolus multiflorus*) plants. Sixteen day-old plants 'decapitated' five days before photograph taken. From the left: buds subject to correlative inhibition in intact plant; buds growing out in decapitated plant to which no hormone was added; buds inhibited in plants treated with 0.1 per cent IAA in lanolin to cut end of stem; rapid outgrowth of buds of decapitated plants treated with gibberellic acid (GA₃) at the cut end of stem.

Currently, research workers in this field consider that auxin production in the apical bud and its movement down the stem in some way induces a flow of nutrients towards the stem apex. This concept, known as the *nutrient-diversion theory*, was first evolved by Went in 1936. Not until 1962 was factual supporting evidence presented by Wareing and co-workers, who found that radioactive nutrients such as $^{32}$P and $^{14}$C-Sucrose moved towards, and accumulated in, regions of high auxin concentration (Fig. 2.11). This process, known as

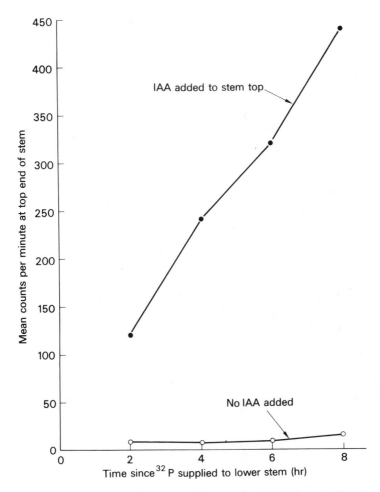

*Figure 2.11.* Effect of auxin (IAA) on the movement of metabolites in decapitated pea plants ('auxin-directed transport'). Application of IAA to the upper end of the stem induced movement of $^{32}$P to the treated tissues from its place of application as $^{32}$P-orthophosphate to the base of the stem (upper line). When no IAA was applied, very little accumulation of $^{32}$P in the upper stem tissues occurred (lower line).
(From C. R. Davies and P. F. Wareing, *Planta* (*Berl.*) **65**, 139–156, 1965.)

*auxin-directed transport*, has been confirmed and amplified in succeeding years, so that it appears that the apical bud receives a preferential supply of nutrients because it is a region of high auxin concentration, and not just because it is a nutrient sink. This explains why exogenous auxin can duplicate the effect of the apical bud in maintaining starvation and correlative inhibition of axillary buds. Other growth hormones, such as gibberellins and cytokinins, do not influence nutrient movement in the way that auxins do, although a mixture of

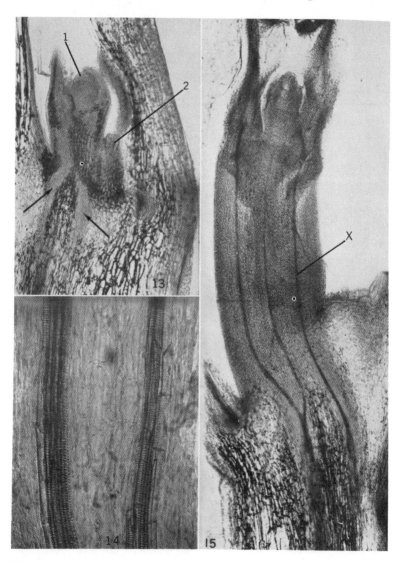

auxin and gibberellin or cytokinin is more effective than auxin alone.

The mechanism of auxin-directed transport has not yet been fully worked out. The most likely explanation is offered by histological studies which have shown that inhibited axillary buds lack properly developed vascular connections with the vascular system of the main stem (Fig. 2.12). This, in itself, suggests that axillary buds receive a poor supply of metabolites. Removal of the apical bud leads to rapid development of vascular connections between previously inhibited axillary buds and the stem, and this development *precedes* visible outgrowth of the buds (Fig. 2.12). Application of an auxin to the top of a decapitated stem prevents both vascular development in the bud and its growth. Thus, basipetally moving auxin in the main stem inhibits the formation of vascular tissue between axillary buds and the main vascular supply of the plant, which perhaps explains why auxin synthesis in the apical bud results in movement of metabolites to that region of the shoot and consequent arrested development of axillary buds.

Information is accumulating which indicates that cytokinins, as well as auxins, are concerned in apical dominance. In contrast to auxins, cytokinins stimulate the outgrowth of buds subject to correlative inhibition. Addition of kinetin to axillary buds of various species (*Pisum, Coleus, Scabiosa, Helianthus annuus, H. tuberosus, Vicia faba*) caused them to grow out even though the apical bud was still present (Fig. 2.13). However, kinetin provided only a temporary release from correlative inhibition, for the buds ceased growth after

---

*Figure 2.12.* Inhibition by auxin, and stimulation by cytokinin, of the formation of vascular connections between axillary buds and the vascular system of the stem in pea seedlings. Stem segments plus buds were incubated in 2 per cent sucrose solutions containing auxin (0.25 mg/l naphthalene acetic acid, NAA) in '13'; in 2 per cent sucrose plus NAA plus also 8 mg/l kinetin in '14'; and in 2 per cent sucrose plus NAA plus 5 mg/l kinetin in '15'. Note that in '13' (auxin but no cytokinin) the bud traces (arrows) to the two buds (1 and 2) have not differentiated into xylem elements. In '14', a more highly magnified view of bud traces shows typical xylem elements developed in the presence of kinetin. The photograph labeled '15' shows the perfect xylem connections which were established between bud and internodal xylem during 72 hr incubation in NAA plus kinetin.
(From Helen P. Sorokin and K. V. Thimann, *Protoplasma* **59**, 326–350, 1965. Original print supplied by Professor K. V. Thimann.)

elongating by one or two centimeters. In 1967, T. Sachs and K. V. Thimann discovered that axillary buds would go on growing following kinetin treatment provided that IAA was also applied (Fig. 2.13). Lack of growth in axillary buds therefore appears to be attributable to them being deficient in both cytokinin and auxin. It is possible that active growth in all shoot meristems requires a supply of cytokinins from the roots, and that the apical bud monopolizes this supply by the process of *auxin-directed transport.*

**(a)**        **(b)**        **(c)**        **(d)**

*Figure 2.13*. Release of axillary buds from correlative inhibition in intact plants by application of a cytokinin, and subsequent treatment of the growing bud apex with auxin.

Left to right: (a), untreated bud; (b), bud treated with kinetin only; (c), bud treated with kinetin followed 3 days later by gibberellic acid; (d), treated with kinetin followed 3 days later by indole-3-acetic acid (IAA). Only buds treated with IAA following kinetin treatment continued growing more or less normally. Gibberellic acid treatment did not effect continued bud growth following kinetin treatment. All hormones were applied directly to the buds dispersed in a mixture of 50 per cent ethanol and 8 per cent carbowax to aid their penetration of the tissues. Concentrations used were; 330 ppm kinetin, 100 ppm $GA_3$, and 1000 ppm IAA. The photograph was taken 7 days after the first hormone treatment. (From Tsui Sachs and K. V. Thimann, *Amer. J. Bot.* **54**, 136–144, 1967. Original print supplied by Professor K. V. Thimann.)

Gibberellins may participate in apical dominance by interacting with auxin from the apical bud, enhancing the auxin-directed transport phenomenon. They do not, however, appear to play any more direct role. Thus, gibberellins do not induce growth in inhibited lateral buds, nor do they substitute for the apical bud, as auxins do, in maintenance of correlative inhibition in decapitated plants (Fig. 2.10).

A few workers have reported that endogenous growth inhibitors, perhaps including ABA, are present in higher concentrations in correlatively inhibited lateral buds than in those released from inhibition by decapitation of the shoot. Much more work is required, nevertheless, before one can reasonably propose that these substances play any part in apical dominance. Similarly, we need to know a lot more than we do about ethylene in relation to apical dominance. There has been one report that ethylene is synthesized more rapidly at the base of a correlatively inhibited bud than at the base of a bud released from inhibition. Work will undoubtedly proceed to evaluate the significance of this observation.

## Root growth

The internal mechanism which controls root growth is less well defined than that which operates in the shoot. By analogy with the situation in coleoptiles and stems (see above), one might expect that auxins and gibberellins are synthesized in the apical part of the root, and that these hormones are translocated basipetally to the elongating zone a few millimeters behind and there regulate extension growth. Similarly, it might be supposed that the vascular cambium of the root is controlled by auxin and gibberellin emanating from the root tip. However, our current state of knowledge of the physiology of the root growth does not allow us to assume that stems and roots possess similar growth-control systems.

### Root elongation

Whereas excision of a stem apical bud, or coleoptile tip, results in a marked reduction in elongation of the lower regions, removal of a root apex does not always cause an inhibition of elongation growth in the remainder of the organ. Some early workers found quite the opposite to be the case, in that a transitory *increase* in the elongation rate of roots occurred when the root tip was cut off. At first sight this result suggests that the root tip exerts only an inhibitory effect on extension growth of cells in the elongation zone, but it appears that this is not necessarily true. For many years it has been accepted that the root tip is a region of auxin synthesis, and we have more recently discovered that both gibberellins and cytokinins are also synthesized at the root apex. Evidence for gibberellin and cytokinin biosynthesis in the root tip is fairly conclusive, but the idea that auxin is also produced there is less soundly established. This does not mean that auxin is not synthesized in root tips, but we must maintain open minds on the matter until further research is done. Recent work on the translocation of IAA has shown that this auxin is transported in a polar manner in roots, *towards* and right up into the root tip (p. 19, Fig. 1.12). Consequently, if IAA is the principal auxin in roots, as it seems to be in shoots, then it is difficult to see how it can be produced in the tip region and serve as a hormone in the more basal parts of the root. It is possible, of course, that a completely different 'root auxin' of some sort is synthesized in the root tip, whose transport is basipetal rather than acropetal, but there is no evidence for this. Isolated root segments of some species have been found to elongate in response to auxins in a manner similar to that seen with coleoptile or stem segments (Fig. 2.14), which indicates that there is no necessity to postulate the existence of a special 'root auxin'.

Root tissues do appear to be particularly sensitive to auxins, and a comparison of Figs. 2.4 and 2.14 shows that although isolated coleoptile, stem and root segments all exhibit elongation responses to exogenous IAA, the optimal concentration of the latter is approximately 10 mg/l for coleoptile and stem elongation but only $10^{-4}$ mg/l for root elongation. In other words, roots are about 100 000 times more sensitive to exogenous IAA than coleoptiles or stems.

In summary, there is considerable *indirect* evidence that IAA is synthesized in the root tip. Also, it has been found that enzyme preparations from lentil (*Lens culinaris*) roots are able to bring about the conversion of tryptophan to IAA, although it is possible that epiphytic bacteria were present in the enzyme preparation (see p. 12). The rather recent findings that exogenous IAA moves acropetally in roots (p. 19, Fig. 1.12) throws long-standing ideas into the melting pot, and the whole vexed question of the role of auxin in root-extension growth requires re-investigation.

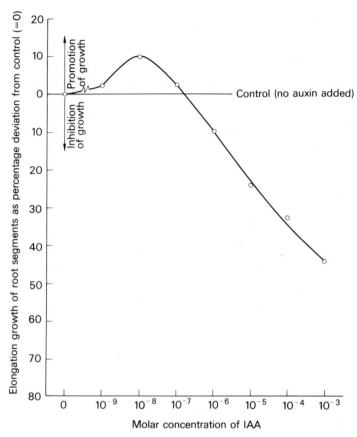

*Figure 2.14.* Elongation growth of sub-apical root segments of *Lens culinaris* in response to exogenous auxin (IAA). Note that only a slight promotion of elongation occurred, and that the optimal IAA concentration was as low as $10^{-8}$ M ($1.75 \times 10^{-3}$ mg/l). Concentrations of IAA greater than $10^{-7}$ M inhibited growth below the control value. (From P. E. Pilet, M. Kobr, and P. A. Siegenthaler, *Rév. Gen. de Botanique*, **67**, 573–601, 1960.)

Despite good evidence showing that gibberellins and cytokinins are produced in root tips, we do not know what part, if any, these growth hormones play in the normal control of root elongation. In contrast to their usually marked effects in aerial parts of plants, applications of exogenous gibberellins have very little effect on root growth in intact plants. Excised roots growing in sterile nutrient cultures sometimes grow more in length when supplied with a gibberellin. Also, whereas excised roots in culture usually fail to form root hairs, they have been reported to do so in the presence of gibberellic acid. Such scattered observations lead one to believe that we shall eventually find that endogenous gibberellins are concerned with the control of root-growth processes. Cultured excised roots also respond to exogenous cytokinins. Kinetin induces an accelerated cell-division rate in the apex of excised pea roots, and this leads to an increase in root vascular tissues.

**Radial growth in roots**

Almost all the increase in diameter which occurs in older regions of roots is a result of the activity of the root vascular cambium. As in shoots, the vascular cambium in roots of dicotyledons is both *fasicular* and *interfasicular* in secondarily thickened regions. It similarly gives rise to xylem and phloem tissues. Much of the evidence which implicates growth hormones in the control of vascular cambial activity in roots has come from studies of excised roots growing in sterile culture. In excised roots, such as those of pea (*Pisum sativum*) or radish (*Raphanus sativus*), only the primary vascular structure develops, even when auxin is supplied in the bathing nutrient medium. When, however, auxin is supplied only through the basal cut end of an excised root, then secondary thickening occurs. The addition of a cytokinin greatly increases the stimulatory effect of auxin on vascular development in excised roots.

In temperate zone trees, cambial activity in roots does not start until the wave of renewed cambial activity in the shoot reaches the base of the trunk. The wave of activity passes into the roots, so that cambial activity progresses acropetally in roots. This suggests that secondary thickening in roots is controlled by auxin entering from the shoot system.

**Root initiation**

Auxin is required for root-primordium initiation. Evidence in support of this statement has come principally from studies of adventitious root initiation in shoot cuttings, and from observations of regeneration in sterile callus cultures.

In the common horticultural technique of vegetative propagation of shoot cuttings, it is normally the case that new roots are formed at the basal end of the stem (Fig. 2.15). It is in just this region that endogenous auxin, which originates in growing leaves or buds on the cutting, would be expected to accumulate as a consequence of the polarized basipetal flux of auxin in stem tissues (p. 17). Cuttings of some species root much less readily than others, but even these may sometimes be induced to root by treatment with an exogenous auxin (known commercially as 'rooting hormones'). We have already seen (chapter 1, p. 32), that shoot buds and root primordia may be caused to differentiate in callus cultures of plant cells by exposure to an appropriate combination of auxin and cytokinin. We may reasonably assume, therefore, that both of these growth hormones are required for the organization of a new root apical meristem. Adventitious root initiation in shoot cuttings is markedly *inhibited* by gibberellins, but the reasons for this are not yet known.

*Figure 2.15.* Induction of adventitious ·root formation by auxins in Western Red Cedar (*Thuja placata*) shoot cuttings. Auxin-treated cuttings were dipped in a mixture of two synthetic auxins, 0.5 per cent naphthalene acetic acid and 0.5 per cent indole-3-butyric acid, dissolved in methyl alcohol, immediately before planting. Photograph taken 7 weeks later.
(Original print supplied by Dr. K. A. Longman.)

Lateral root initiation has mainly been studied in excised roots growing in sterile nutrient culture. In general, the addition of an auxin to the medium promotes the formation of lateral roots. However, factors other than auxin also appear to be required for lateral root initiation in excised roots. These other factors are partially nutritional, but also include unidentified substances which are supplied by older root tissues. Cytokinins are also important for lateral root initiation. Thus, Torrey and others have noted that auxin-induced lateral root production in isolated pea-root segments is enhanced by a supply of a certain concentration of kinetin or adenine. It appears that lateral root initiation is at least partially determined by the relative concentrations of auxin and cytokinin present, in a similar manner to root initiation in callus cultures (p. 32). In contrast to the inhibitory effect of gibberellins on adventitious root initiation, it has been reported that, under certain conditions, gibberellic acid can promote the formation of lateral roots in excised tomato roots.

Finally, it should be mentioned that the root apical meristem exerts an inhibitory effect on lateral root initiation, a situation reminiscent of the correlative inhibition of axillary buds by the apical bud in shoots. Thus, excision of the root tip leads to an increased rate of lateral root initiation, and also to an increase in the absolute number of lateral roots formed. However, we have no good reason to believe that suppression of lateral root initiation by the root apex is a result of auxin synthesis in the latter. Indeed, there is no unequivocal evidence that auxin is synthesized in the root tip (p. 19) and auxin transport in roots is acropetal, the opposite situation to that pertaining in the shoot (p. 19). Also, of course, auxin enhances rather than suppresses root initiation.

## Further reading list

1. Heslop-Harrison, J. 'Plant Growth Substances', in *Vistas in Botany* Vol. 3. *Recent Researches in Plant Physiology* (ed. W. B. Turrill), pp. 104–194, Pergamon Press, London, 1963.
2. Leopold, A. C. *Plant Growth and Development*, McGraw-Hill, N.Y., 1964.
3. Phillips, I. D. J. 'Apical Dominance', in *The Physiology of Plant Growth and Development* (ed. M. B. Wilkins), pp. 165–202, McGraw-Hill, London, 1969.
4. Street, H. E. 'The Physiology of Root Growth', *Ann. Rev. Plant Physiol.* **17**, 315–344, 1966.
5. Wareing, P. F., C. E. A. Hanney, and J. Digby. 'The Role of Endogenous Hormones in Cambial Activity and Xylem Differentiation', in *The Formation of Wood in Forest Trees* (ed. M. Zimmerman), Academic Press, N.Y., 1964.
6. Wareing, P. F. and I. D. J. Phillips. Chapter 5 in *The Control of Growth and Differentiation in Plants*, Pergamon Press, Oxford, 1970.

# 3. Growth hormones in phototropism and geotropism

Multicellular plants differ from multicellular animals in many ways, but to the layman the most obvious distinction between them is that animals can move around, whereas plants are fixed in one position. Although there are exceptions to this rule, as a general statement it is true, and is undoubtedly a reflection of differing modes of nutrition; the majority of animals have to forage for their food, but most higher plants are autotrophic, manufacturing organic substances from inorganic materials in their immediate environment. Certain aquatic algae, and male reproductive cells of higher plants other than those of angiosperms, quite clearly do move as a result of the action of flagella or cilia. Apart from these cases of locomotory plant cells, however, the capacity for movement in plants is restricted to individual organs of the sedentary whole organism.

Various types of movement of organs occur in higher plants, some of which take place as a result of growth activities. The direction of extension growth in shoots and roots is determined by several components of the environment, particularly gravity and light. Thus, organs of the shoot- and root-system become orientated in response to *directional* gravitational and light stimuli. Movements in plants that occur in response to directional external stimuli are called *tropisms*, and since these always involve growth they are examples of *growth movements*. Growth movements are invariably brought about by unequal, or differential, extension growth in opposite sides of an extending organ, which causes the organ to bend in a particular direction. Consequently, it is mainly in younger regions

**73**

of the plant that growth movements are executed, although older parts that still retain the capacity for extension growth may also show limited growth movements.

In some plants, bending movements take place which are not dependent upon growth, but are a result of changes in turgor of certain specialized regions known as pulvini. For example, variation, or sleep, movements of leaves such as those of clover (*Trifolium*) take place over and over again, due to diurnal alterations in the water content of the pulvinus situated at the base of each leaflet. Similarly, the rapid coiling responses shown by tendrils of some plants when stimulated by various physical or chemical factors (thigmotropism and thigmonastisism), occur as a result of rapid turgor changes in cells of the tendril (for a full modern account of tendril physiology, see M. J. Jaffe and A. W. Galston, *Ann. Rev. Plant Physiol.* **19**, 417–434, 1968).

However, we are concerned here with growth movements, for these have received intensive study which has revealed that auxin, and perhaps other growth hormones, are involved in their occurrence. The phenomena of *phototropism* and *geotropism* are by far the best-studied examples of growth movements in plants. Other tropisms also exist, but the mechanisms of these are much less clear to us at present. There is, in any case, no doubt that the orientation of different parts of a plant with respect to light and gravity is of considerable importance in its normal development.

## Phototropism

This term is applied to a growth movement that occurs in response to a directional light stimulus. When plants are exposed to light coming more strongly from one direction than others (*unilateral illumination*), then individual organs may perform *phototropic curvatures* which lead them to reach an equilibrium position in relation to the direction of incident light.

Phototropically responding plant organs may bend either toward a light source or away from it. When the movement is toward the light source it is said to be *positive phototropism*, and when away *negative phototropism*. Most aerial regions of higher plants are positively phototropic, whereas roots and other underground organs are often negatively phototropic. Nevertheless, quite a number of exceptions to these rules exist, and, in addition, the direction of response to unilateral illumination in any given organ may be reversed as it gets older, or when the intensity of incident light is altered.

**Phototropism in coleoptiles**

Phototropism in etiolated coleoptiles of grass and cereal seedlings has been studied far more than phototropism in any other organs or plants. Dark-grown coleoptiles are very sensitive to unilateral illumination, curving rapidly by differential growth rates in the illuminated and shaded sides. The response may be positively phototropic or negatively phototropic, depending principally on the total *quantity* of unilateral light received, i.e., the intensity of the light multiplied by the period of exposure (Fig. 3.1). Most studies

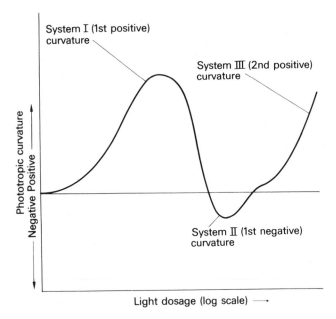

*Figure 3.1.* Schematic summary of Systems I, II, and III phototropic curvatures of oat coleoptiles exposed to monochromatic blue light. Coleoptiles of other cereals show a similarly shaped curve, except that System II may not be clearly negative.

(Adapted from Winslow R. Briggs, chapter 8, pp. 223–271, in *Photophysiology* Vol. I (ed. Arthur C. Giese), Academic Press, N.Y. and London, 1964.)

have investigated positive phototropism in etiolated coleoptiles, induced with relatively low dosages of unilateral light—i.e., *System I* or *1st positive curvature* (see Fig. 3.1).

Not all wavelengths of light induce phototropic responses in coleoptiles. Figure 3.2 shows the action spectrum for positive phototropism in the oat (*Avena*) coleoptile, and it can be seen that blue light is more effective than other regions of the visible spectrum. We must therefore assume that a natural pigment, or pigments, absorb radiation of active wavelengths, the energy of which is then utilized in the phototropic mechanism. The important pigments involved appear to be riboflavin and $\beta$-carotene, both of which absorb light of the blue wavelengths (Fig. 3.2).

As we saw in chapter 1 (p. 6), Charles Darwin demonstrated that unilateral illumination is perceived in the tip region of a coleoptile, but that the growth curvature response occurs lower down in the region of extension growth (Fig. 1.1). This important point was confirmed by other workers, and in 1911 Boysen-Jensen found that a stimulated coleoptile tip transmits a chemical, later isolated and called auxin (p. 7), which induces the growth curvature in lower parts of the coleoptile (Fig. 1.1). Cholodny and Went, two plant physiologists working independently, suggested in the late 1920's that under the influence of unilateral light auxin migrates *laterally*, away from the light source. This, the *Cholodny–Went theory* of phototropism, therefore suggested that differential growth in a phototropically responding coleoptile occurs as a consequence of a difference in auxin concentration between the light and dark halves.

Boysen-Jensen performed a simple experiment with oat coleoptiles, in which a mica sheet was inserted across half the cross-sectional area just below the tip to serve as a barrier to auxin flow. When the barrier was below the illuminated half of the tip a normal positive phototropic curvature developed, but a barrier positioned below the darkened half of the tip prevented curvature (Fig. 3.3a). This suggested that phototropism involves the transmission of auxin from the less brightly illuminated portion of a coleoptile tip, but it did not provide any evidence for lateral displacement of auxin molecules from the light to dark sides as suggested by Cholodny and Went. However, later but similar experiments by Boysen-Jensen did support the Cholodny–Went theory of phototropism. In these, a barrier of mica or thin glass sheet was inserted vertically into

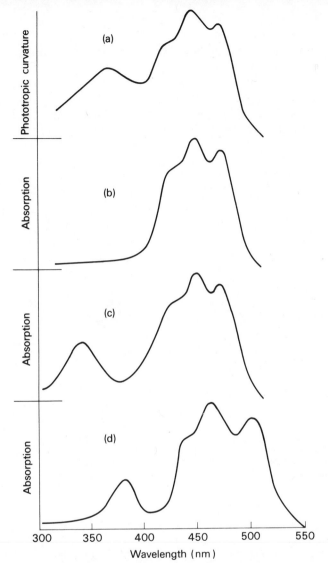

*Figure 3.2.* A comparison of the action spectrum (a) for oat coleoptile phototropic curvature (System I) with the absorption spectra of *trans-β*-carotene (b), *cis-β*-carotene (c), and riboflavin (d).

None of the three pigments exactly qualifies as the phototropism photoreceptor. However, it is thought that riboflavin is the most likely candidate, for its absorption spectrum (d) can be shifted to the left to correspond with the phototropic action spectrum (a) by dispersion of the pigment in a suitable lipoidal solvent.

(From M. Bara and A. W. Galston, *Physiol. Plant.* **21**, 109–118, 1968.)

coleoptile tips. The coleoptiles were then exposed to unilateral light arriving either parallel to or at right angles to the barriers (Fig. 3.3b). A positive phototropic curvature ensued only when the barrier was parallel to the light rays. It appears from these experiments that a growth curvature occurs only when lateral migration of auxin from light to dark halves is not impeded by a barrier.

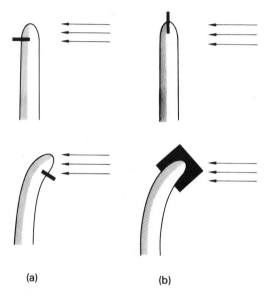

(a)                              (b)

*Figure 3.3*. Two experiments performed by Boysen-Jensen in 1928, which indicated that auxin moves preferentially down the shaded half of a unilaterally illuminated coleoptile (a), and that this involves lateral transport of auxin from the illuminated to the shaded side (b). Only when the route of basipetal auxin movement in the darkened side was unimpeded did a growth curvature appear (a). Similarly, a mica sheet inserted vertically into the tip region at right angles to incident light prevented a phototropic response, presumably because auxin could not migrate laterally. A similar mica sheet inserted parallel to the light rays did not prevent curvature (b).

More direct evidence for lateral auxin transport in phototropically stimulated coleoptile tips was provided by Went in 1928, who found that greater quantities of diffusible auxin (see p. 10) could be collected in agar from below the shaded halves of oat coleoptile tips, than from the illuminated halves of the same tips. Went deduced that the lateral differential in auxin concentration which is induced by unilateral illumination can only be explained in terms of lateral auxin transport. Nevertheless, for reasons we need not consider here, Went's results did not provide completely unequivocal evidence in support of the Cholodny–Went theory. Because of this, several other theories, all invoking different rates of auxin synthesis or destruction in illuminated and darkened tissues, were put forward to explain the occurrence of lateral differentials in auxin concentration in phototropically responding organs. However, it is currently considered that lateral transport of auxin *does* afford a complete explanation for differences in auxin concentration between light and dark regions of unilaterally illuminated etiolated coleoptiles. Convincing evidence for this view has come from studies of the distribution of endogenous auxin (Fig. 3.4) and of the lateral displacement of exogenous radioactive auxin ($^{14}$C-IAA), following phototropic stimulation (Fig. 3.5).

In coleoptiles, therefore, auxin synthesis goes on in the tip region both in complete darkness or in light. When light of a dosage which induces a positive phototropic curvative is shone onto a coleoptile tip from one side, auxin produced in the tip is displaced toward the shaded flank of the coleoptile. This causes a higher concentration of auxin to be present in the shaded side, which consequently elongates more rapidly than the illuminated side—i.e., a positive phototropic curvature develops. While we can feel confident that this is a satisfactory description of what occurs, it must be remembered that the effect of unilateral light upon the pattern of auxin transport is itself a result of the activation of some perception mechanism in the tip. That is, auxin molecules are not themselves sensitive to differences in light intensity.

The mechanism by which light energy is utilized in the lateral transport of auxin is completely obscure at the present time. An attractive hypothesis, which was put forward some years ago, suggested that an electrical potential difference between light and dark sides of a unilaterally illuminated organ is involved. Such *bioelectrical potentials* occur spontaneously in phototropically responding organs, and it was suggested that ionized acidic auxin molecules migrate electrophoretically toward the more electrically positive dark side. Lateral potential differences of as much as 100 mV have been measured in unilaterally illuminated coleoptiles, but it has been discovered in recent years these do not occur until *after* lateral auxin transport has already started. Thus, lateral bioelectric potentials in phototropically responding plant organs cannot be considered to be the cause of lateral auxin transport.

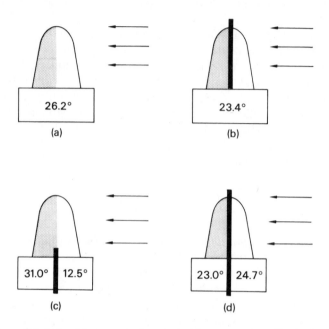

*Figure 3.4.* Lateral transport of auxin in *Zea mays* coleoptile tips in response to unilateral illumination from the direction indicated by arrows. The tips were stood on agar blocks for 3 hr and received 125 000 meter-candle-seconds of light from one side. The figures indicate the amount of auxin which diffused into each agar block, expressed as degrees curvature induced by the blocks in the *Avena* curvature test (see Fig. 1.2). The dark vertical lines represent thin glass barriers positioned at right angles to incident light rays.

(a) Intact coleoptile tip; (b) tip bisected but agar block not, with no significant effect on total auxin yield from the tip; (c) agar block bisected, which resulted in more auxin diffusing into the block below the shaded half of the tip; (d) tip and agar block bisected, when approximately equal quantities of auxin diffused into the two agar blocks.

Approximately twice the total auxin yield of (a) and (b) was obtained in (c) and (d), because in the latter each agar block had been in contact with six half-tips, the equivalent of three whole tips which were placed on agar blocks in (a) and (b).

(From W. R. Briggs, R. D. Tocher, and J. F. Wilson, *Science* **126**, 210–212, 1957.)

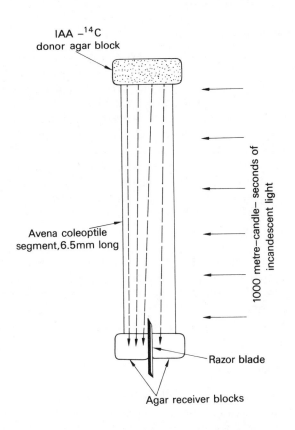

IAA –$^{14}$C
donor agar block

1000 metre–candle– seconds of
incandescent light

Avena coleoptile
segment, 6.5mm long

Razor blade

Agar receiver blocks

| Experiment number | Counts/min in receivers | | Ratio of auxin collected (shaded : illuminated) |
| | Below illuminated side | Below shaded side | |
| --- | --- | --- | --- |
| 1 | 10.9 | 28.4 | 72 : 28 |
| 2 | 14.3 | 50.4 | 78 : 22 |
| 3 | 15.2 | 50.6 | 77 : 23 |
| 4 | 22.1 | 61.9 | 74 : 26 |
| 5 | 39.0 | 66.2 | 63 : 37 |
| 6 | 35.2 | 82.0 | 70 : 30 |
| 7 | 48.3 | 158.0 | 77 : 23 |
| 8 | 91.0 | 170.0 | 65 : 35 |
| 9 | 118.0 | 157.0 | 66 : 34 |
| Means | 43.8 counts/min | 91.6 counts/min | 71.4 : 28.6 |

*Figure 3.5.* Lateral displacement of radioactive indole-3-acetic acid
(IAA-$^{14}$C) in a coleoptile segment as a result of unilateral illumination.
Radioactivity appearing in the two agar receiver blocks was measured
separately and expressed as counts per minute above background.
(From Barbara Gillespie Pickard and K. V. Thimann, *Plant Physiol.* **39**,
341–350, 1964.)

## Phototropism in green shoots

Leafy dicotyledonous green shoots are clearly more complex structures than etiolated coleoptiles. They are nevertheless sensitive to unilateral illumination. In some cases, particularly in plants such as sunflower (*Helianthus annuus*), they exhibit a form of phototropism called *heliotropism*. This term is applied to 'sun-following' growth movements, whereby the apical part of the shoot bends toward the sun as the latter moves across the sky each day (Fig. 3.6). Some work has demonstrated that heliotropic movements are related to asymmetry of auxin distribution in the extending region of the stem, and that this in turn results from differences in the rates of auxin synthesis in shaded and brightly illuminated leaves on opposite sides of the stem (Fig. 3.7).

Leafy shoots of dicotyledonous plants are not only heliotropic, but they also have a phototropic mechanism probably basically similar to that of etiolated coleoptiles, but this has received very much less study than that in etiolated coleoptiles. However, we do know that lateral transport of auxin can occur in green dicotyledonous stems as well as in coleoptiles, suggesting that they have similar phototropic mechanisms.

## Negative phototropism

Relatively little work has been performed on the relationship between auxin and negative phototropic curvatures (i.e., curvature away from the light source). As stated at the beginning of this chapter (p. 74), negative phototropism occurs particularly in subterranean plant organs, but also in aerial organs exposed to particular dosages of unilateral light (see Fig. 3.1). The mechanism of negative phototropism in underground organs such as roots has hardly been studied at all. The distribution of auxin in 'system II' (Fig. 3.1) negative phototropically responding etiolated coleoptiles has been found to be correlated with the direction of curvature—i.e., a higher auxin concentration is present in the more rapidly elongating

*Figure 3.6*. Heliotropic growth movements in a single sunflower (*Helianthus annuus*) plant photographed from the north side at different times of the same day.
(From H. Shibaoka and T. Yamaki, *Sci. Papers Coll. Gen. Education, University of Tokyo*, **9**, 105–126, 1959. Original prints supplied by Dr. Hiroh Shibaoka.)

Morning

Noon

Evening

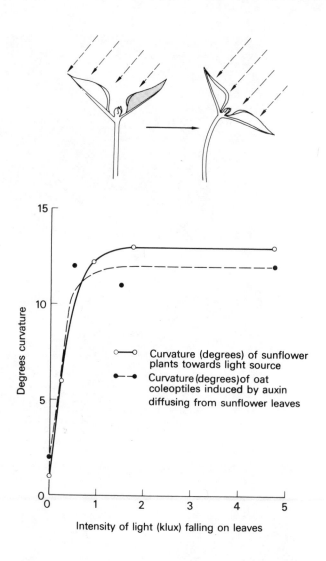

*Figure 3.7.* Relationship between auxin production by opposite leaves and heliotropic growth curvature in the stem of *Helianthus annuus* plants. *Above.* When light is not falling evenly on each member of a pair of leaves, a heliotropic curvature takes place until both leaves are equally illuminated. *Below.* The effect of light intensity on auxin production in leaves, and on induction of heliotropic growth curvatures.

(From H. Shibaoka and T. Yamaki, *Sci. Papers Coll. Gen. Education, University of Tokyo*, **9**, 105–126, 1959.)

illuminated region. Thus, with a light dosage sufficiently high to induce a system II (negative) phototropic curvature, lateral auxin transport is toward the light source, rather than away from it (Table 3.1). Just as we do not yet understand the mechanism of lateral auxin transport in positive phototropism, so we have no explanation of the direction of auxin displacement in negative phototropically responding coleoptiles.

**Table 3.1**

Correlation between direction of phototropic curvature and amounts of auxin diffusing from illuminated and shaded halves of the uppermost 2 mm of oat coleoptile tips after unilateral illumination. (From M. Wilden, *Planta* (*Berl.*) **30**, 286–288, 1939–40).

| Quantity of light given meter candles × seconds | Nature of curvature | Ratio of auxin diffused (illuminated : shaded)* |
|---|---|---|
| 100 × 15 | positive | 17:83 |
| 380 × 30 | negative | 62:38 |
| 3,000 × 50 | positive | 36:64 |

* Based on curvatures induced by agar blocks in Went *Avena* test.

## Geotropism

In common with other tropistic responses in plants, the stimulus which occasions the appearance of a geotropic curvature is a directional stimulus; the gravitational field acting toward the center of the earth. Plant organs exhibit several forms of geotropism. They may show *negative geotropism* (grow away from the earth's center; e.g., coleoptiles and stems), *positive geotropism* (grow toward the center of the earth; i.e., roots, particularly primary roots), *diageotropism* (grow horizontally at right angles to the gravitational field; e.g., rhizomes and stolons), or *plagiogeotropism* (grow at an angle other than 90° to the gravitational field; e.g., lateral branches, leaves, and some roots).

Geotropism is similar to phototropism in several respects: (a) it is a response to a directional stimulus, (b) it occurs as a result of differential growth leading to curvature of an elongating organ, (c) perception of the stimulus takes place in the apical region of coleoptile, shoot or root and, (d) transmission of a stimulus occurs from the region of perception to the region of response.

Clearly, therefore, the apical parts of shoots and roots contain a geoperception mechanism. Histological time-course studies have revealed that when an organ such as a coleoptile or root is displaced from its normal, or preferred, vertical orientation to a horizontal position, amyloplasts move across certain cells (*statocytes*) and come to lie along what were previously vertical walls of the cells. There are good reasons for regarding the movement of these amyloplasts (or, in the context of geoperception, what may be called *statoliths*) as representing the mechanism of gravity perception in plants. We are not at all clear what effects the re-distribution of amyloplasts have upon the cell's internal environment, but it is only following sedimentation of amyloplasts in cells of the tip region that transmission of a stimulus to elongating regions ensues, with the consequent development of a geotropic growth curvature. In roots, the starch grain-rich cells of the root cap appear to play an important role in geoperception. Removal of the cap from *Zea mays* roots destroys sensitivity to gravity without preventing straight extension growth. Not until a new cap regenerates does geotropic behaviour reappear.

Just as in phototropism, there is a correlation between the differential elongation growth which leads to a growth curvature in geotropically responding organs, and auxin concentrations. Higher levels of auxin appear in tissues of the *lower* sides of coleoptiles, stems, and roots, when these organs are moved to a horizontal position. It is generally considered that the auxin is produced in the tip region, and that as a result of geotropic stimulation lateral auxin transport occurs in addition to normal basipetal transport of auxin (Fig. 3.8). In the case of coleoptiles and stems, a negative (upward) curvature develops due to more rapid cell elongation in the lower than in the upper side of the organ, owing, perhaps, to the stimulatory effect of the higher concentration of auxin in the lower tissues.

Moving a root from its normal vertical (downward pointing) to a horizontal position causes a positive (downward) growth curvature to develop as a result of more rapid elongation in the *upper* side of the elongating zone. Measurements of auxin levels have revealed, nevertheless, that similarly to what occurs in coleoptiles and stems, the auxin concentration in a horizontal root is greater in the tissues of the lower than the upper side. Thus, the more rapidly elongating upper tissues in geotropically curving roots contain lesser auxin levels than the slower growing lower tissues. It is not possible at present to give a completely adequate explanation for this apparent

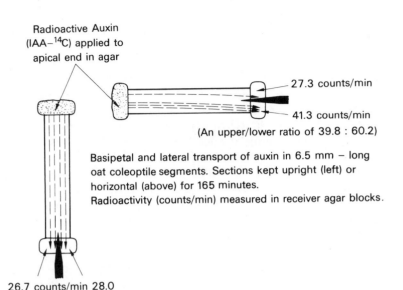

Radioactive Auxin (IAA–$^{14}$C) applied to apical end in agar

27.3 counts/min

41.3 counts/min

(An upper/lower ratio of 39.8 : 60.2)

Basipetal and lateral transport of auxin in 6.5 mm – long oat coleoptile segments. Sections kept upright (left) or horizontal (above) for 165 minutes.
Radioactivity (counts/min) measured in receiver agar blocks.

26.7 counts/min 28.0
(A ratio of 49.1 : 50.9)

*Figure 3.8.* Lateral displacement of radioactive indole-3-acetic acid (IAA-$^{14}$C) in a coleoptile segment induced by a transverse gravitational stimulus. Radioactive IAA collected in agar receiver blocks.
(From Barbara Gillespie Pickard and K. V. Thimann, *Plant Physiol.* **39**, 341–350, 1964.)

paradox. The usual reason given in textbooks involves the greater sensitivity to auxin of roots as compared with shoots (Figs. 2.4 and 2.14). Thus, it is envisaged that the accumulation of auxin in the lower side of coleoptiles and stems proceeds to the extent that an optimal concentration for the elongation of the tissues is achieved. In contrast, similar accumulation of auxin in the lower tissues of a horizontal root might be expected to be supra-optimal and inhibitory to elongation, with consequently more rapid growth on the upper side. This view may be correct, but current confusion as to the role of auxin in root elongation growth, and the problem of whether or not auxin is synthesized in root tips (p. 68), means that we must keep open minds on the possibility that positive geotropism in roots is mediated through supra-optimal auxin concentrations in the lowermost tissues. Nevertheless, by analogy with what occurs in coleoptiles and stems, with respect to lateral auxin transport in phototropism and geotropism, and the good experimental evidence for similar lateral displacement of auxin in geotropically stimulated roots, it is reasonable to suppose that auxin is in some way closely involved in positive geotropism in roots.

Very recent work by Wilkins and coworkers, with *Zea mays* roots from which the root cap can be excised, has revealed that when one-half of the cap is removed then the root executes a growth curvature towards the remaining half root cap, regardless of the direction in which gravity is acting on the root. It appears that the root cap secretes a growth inhibitor to the elongating regions of the root, and that gravity can influence the lateral distribution of this inhibitor in the root tissues. Work is proceeding to test this possibility, and to isolate and characterize the growth inhibitor from root caps.

Downward lateral auxin transport in horizontally positioned organs occurs by some as yet unknown mechanism. There is apparently no difference in the velocity of basipetal polar flux of auxin in the tissues of the upper and lower half, so that differentials in auxin concentration are purely the result of downward lateral transport. Similarly to what has been observed in phototropically responding organs (p. 79), a lateral bioelectric potential occurs in gravity-stimulated plant organs. The lower surface becomes positively charged with respect to the upper surface, with a potential difference of 5–20 mV. This *geoelectric potential* appears some ten or fifteen minutes after the organ is moved from a vertical to horizontal position, at which time the first visible signs of a growth curvature response appear. In other words, the potential difference observed in gravity stimulated organs appears to be an incidental phenomenon, rather than a causal factor in geotropism, and there is no reason to believe that lateral auxin transport is induced by an electrical field across a geotropically responding organ. This conclusion is the same as that reached from studies of phototropism (p. 80).

## Other tropisms

We have considered the role of auxin in positive and negative phototropism and geotropism. Very few studies have been made of auxin relationships in tissues of organs which exhibit other types of geotropism (see p. 85), or tropisms of types which are not described in this book. It is, therefore, not possible to usefully evaluate the possible role of auxin in these other types of growth movement.

Similarly, we cannot ascribe any roles for growth hormones other than auxin in plant growth movements. Some largely circumstantial evidence exists which suggests that gibberellins and ethylene may be involved in plagiogeotropism and positive geotropism, but this cannot be regarded as conclusive.

## Further reading list

1. Audus, L. J. 'The Mechanism of the Perception of Gravity by Plants', *Symposia Soc. Exper. Biol.* **16**, 197, 1962.
2. Audus, L. J. 'Geotropism', in *The Physiology of Plant Growth and Development* (ed. M. B. Wilkins), pp. 203–242, McGraw-Hill, London, 1969.
3. Curry, G. M. 'Phototropism', in *The Physiology of Plant Growth and Development* (ed. M. B. Wilkins), pp. 243–273, McGraw-Hill, London, 1969.
4. Thimann, K. V. and G. M. Curry, 'Phototropism and Phototaxis', in *Comparative Biochemistry* (ed. M. Florkin and H. S. Mason), Vol. I, p. 243, Academic Press, New York, 1960.
5. Wareing, P. F. and I. D. J. Phillips. Chapter 7 in *The Control of Growth and Differentiation in Plants*, Pergamon Press, Oxford, 1970.
6. Wilkins, M. B. 'Geotropism', *Ann. Review of Plant Physiology* **17**, 379–408, 1966.

# 4. Hormones and reproduction in higher plants

The initial event of sexual reproduction in higher plants is *floral induction*, or *flower initiation*, and this involves transformation of the stem apical meristem, or apical meristems if the lateral branches also bear flowers, from a structure of indeterminate growth in which vegetative organs are initiated (leaves, vegetative lateral buds, stem tissues) to one of usually determinate growth which gives rise to specific numbers of petals, sepals, stamens and carpels. The transition of an apical meristem from a vegetative apex to a floral apex is a critical stage in the life of seed plants. It can be classified as a *phase change*, marking a profound and relatively sudden alteration in the pattern of development (see chapter 5, p. 106).

Following flower initiation, the newly formed flower parts proceed to grow and differentiate. The end of flower development is usually considered to have been reached when the anthers open to release the pollen grains; a stage of flower development called *anthesis*. Upon successful pollination and formation of a zygote by union of the male and female haploid reproductive nuclei within the ovary, development of the seed and surrounding fruit structures ensues (Fig. 4.1), and this phase of reproduction is usually termed *fruit growth* (though more correctly it should be referred to as *fruit development*, see p. 2).

In summary, reproduction in seed plants occurs as a result of a sequence of complex developmental events; flower initiation, flower development, and fruit development. We will now consider what is known of the parts played by growth hormones in the control of these.

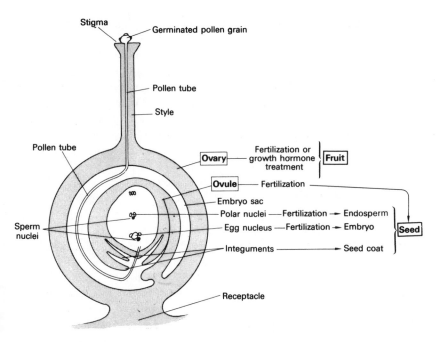

*Figure 4.1.* Fertilization in an angiosperm flower, which normally leads to seed and fruit development. The ovary develops into the fruit tissues, following either fertilization or hormone treatment. The ovule develops into a seed only if fertilization by the two sperm nuclei occurs. Both the egg and polar nuclei are fertilized, and these give rise to the embryo and endosperm respectively.
(Adapted from A. W. Galston, *The Life of the Green Plant*, Prentice-Hall, New Jersey, 1961.)

## Flower initiation

As early as 1880, it was envisaged by a pioneer plant physiologist, Sachs, that the initiation of flowers is under the control of a particular chemical or chemicals within the plant. Sachs had no idea of the nature of the chemical stimulus, of course, and in fact he considered that 'organ-forming substances' existed in plants, each specific for a given organ. We are now aware that the known growth hormones, operating interdependently, fulfil many of the functions of Sachs' hypothetical organ-forming substances (e.g., in stem growth, leaf growth, root and shoot-bud initiation), and it is attractive, though perhaps dangerous, to consider that flower initiation, too, may be prompted by interacting effects of auxins, cytokinins, gibberellins, abscisic acid and ethylene. It is certainly clear that a typical hormonal system ('action at a distance') can operate in controlling the time of flowering, for it has frequently been demonstrated that one region of the plant produces a flower-inducing stimulus, which is transmitted to the tissues of the stem apex, which respond to it by commencing to produce floral parts. This system is most clearly seen in those plants which flower only when grown under a régime with an appropriate proportion of light (*the photoperiod*) in each twenty-four hour cycle. Such species are therefore daylength sensitive, measuring in some manner the relative lengths of light and dark periods to which they are exposed. Some species (e.g., cocklebur, *Xanthium*) flower only under short days (about ten hours or less light every twenty-four hours) and are called short-day plants (SDP) whereas, in contrast, long-day plants (LDP) (e.g., *Hyosycamus niger*, *Perilla*) flower only under long photoperiods (about fifteen or more hours light every twenty-four hours). The perception of prevailing photoperiodic conditions by SDP and LDP involves a special proteinaceous pigment, *phytochrome*. Although the flowering response occurs at the shoot apex, the perception of daylength takes place in the leaves (Fig. 4.2).

When mature leaves of SDP or LDP are maintained under short days or long days, respectively, they become *induced*, and in consequence a stimulus is generated which is transferred to the apical meristem. A great deal of effort has been expended in attempts to identify the nature of the stimulus which arises in photoinduced leaves. The problem has proved singularly intractable, although the circumstantial evidence for the existence of a transmissible flower-inducing hormone is very considerable (sometimes the transmissible floral stimulus is called *florigen*, from the Greek 'flower maker'). No flower-inducing hormone has yet been isolated from plants, but this does not necessarily mean that florigen does not exist. A velocity of 2 to 20 mm/hr for the translocation of the floral stimulus has been measured in several plants (Fig. 4.2) and the direction of its movement within the plant deliberately modified. By grafting techniques it has been shown that florigen can be transferred from a photoperiodically induced plant into a noninduced plant, and there cause

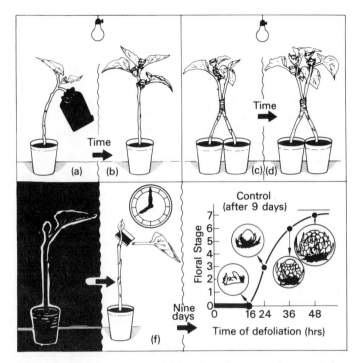

*Figure 4.2.* Experiments with cocklebur (*Xanthium pensylvanicum*) plants, a short-day species, which demonstrate the existence of a flowering hormone (*florigen*). (a)–(b) Exposure of just one leaf to long dark periods results in that leaf becoming induced, and transmission of a flowering stimulus from the leaf to buds. (c)–(d) The flowering stimulus passes through a graft union from an induced plant (left) to elicit a flowering response in the noninduced partner (right). (e)–(f) Measurement of the rate of translocation of the floral stimulus using *Xanthium pensylvanicum* which will flower after exposure to just one long dark period. Removal of the induced leaf immediately after the end of the inductive dark period results in the plant remaining vegetative. If the induced leaf is left attached to the plant for some hours after the end of the 16 hr dark period (see graph), then the plant may flower almost as well as if the leaf were not removed at all.

(From Frank B. Salisbury, *Ann. N.Y. Acad. Sci.* **144**, Art. I, 295–304, 1967.)

flower initiation (Fig. 4.2). Unlike the known growth hormones, the floral stimulus apparently does not pass through nonliving aqueous media such as gelatine or agar-gel, but its translocation depends upon activities of living cells. Because of this, and its low velocity of movement in plant tissues, it has been suggested that florigen is a protein, nucleoprotein, or nucleic acid, which moves from living cell to living cell in a manner similar to the spread of a virus infection. This may be so, but we still lack any sort of conclusive evidence for the chemical nature of the floral stimulus. In addition, one recently reported experiment demonstrated that in *Pharbitis* the floral stimulus can be translocated at a velocity as high as 51 cm/hr, which approaches the velocities of carbohydrate translocation in the phloem.

The weight of evidence is therefore against any idea that the known hormones which are associated with control of growth and differentiation (auxins, cytokinins, gibberellins, abscisic acid, ethylene) are also concerned in the control of flower initiation. However, growth hormones have been shown to modify the normal pattern of flowering in a number of plant species.

## Auxins

Hormones of this type can, provided they are applied at an appropriate time, modify the time of flower initiation in photoperiodically sensitive plants. Only a few species, notably pineapple, regularly show a flowering response to exogenous auxins. Much more commonly observed are inhibitory effects of applied auxin upon flower induction, and this is particularly the case in SDP, although a number of LDP, also, fail to flower under long days when treated with high concentrations of auxins. A number of experiments has been conducted to determine whether endogenous auxin levels in photoperiodically sensitive plants change in response to photoinductive treatment, but the results have been equivocal and contradictory. Thus, available evidence suggests that auxins do not play a direct part in the processes of flower induction.

## Gibberellins

In 1956, Lang discovered that *Hyoscyamus* plants could be induced to flower by treatment with gibberellic acid ($GA_3$). Lang's experiments were performed with biennial *Hyoscyamus niger*, which normally requires a cold treatment (vernalization) before it will subsequently flower at good growing temperatures under long days. Treatment of unvernalized *Hyoscyamus* with $GA_3$ caused flowering to occur, provided the plants were kept under long days. Under short days the shoot elongated but flowers did not form. It has since been repeatedly demonstrated by many workers that exogenous $GA_3$ often induces a flowering response in LDP which are held under nonphotoinductive conditions (i.e., short days) (Fig. 4.3). Does this mean that $GA_3$ is florigen? For a number of reasons, some of which are complex and would consume excessive space if pursued in detail here, the answer to this question appears to be 'no'. One difficulty in the way of assuming that gibberellin is florigen that you may consider, is that gibberellins induce flowering only in LDP and *not* in

*Figure 4.3*. Effect of gibberellic acid on stem elongation and flowering in the long-day plant *Samolus parviflorus*. All plants kept under short days. *Extreme left*. Untreated control. *Second left to extreme right*. Treated with increasing doses of gibberellic acid (1,2,5,10 and 20 μg per plant per day, for approximately three weeks).
(From A. Lang, *Proc. Nat. Acad. Sci. U.S.* **43**, 709–717, 1957. Original print supplied by Professor Anton Lang.)

SDP, yet we know from grafting experiments that the floral stimulus is the same in both, for a noninduced SDP can be caused to flower by grafting it on to an induced LDP, or vice versa (Fig. 4.4).

In general, gibberellins have flower-inducing effects only in rosette plants (see p. 47), or in the so-called 'long-short-day plants', such as *Bryophyllum crenatum* and *B. diagremontianum*, which normally have to be exposed to long days before they flower in response to subsequent short-day treatment. In all these, flower initiation is preceded by sudden and rapid stem extension ('bolting'). When either LDP or long-short-day plants are not exposed to long days, then stem elongation is very considerably less than occurs in long days. In such species, it appears that under short days endogenous

(a)

(b)

*Figure 4.4.* The flowering stimulus (*florigen*) generated in photoinduced leaves is the same in long- and short-day plants. This can be demonstrated by transmission of the stimulus through graft unions between long- and short-day plants. (a) Transmission from an induced long-day plant (*Nicotiana sylvestris*) to noninduced short-day plant (*N. tabacum* cv. Maryland Mammoth). The members of the grafted pair on the right were both kept under long days; long-day partner therefore induced, and short-day partner flowering (flowers of donor long-day plants were removed immediately after their appearance, which increases the response of the grafted partner). The pair of plants on the left were treated

gibberellins are limiting for stem elongation, but that florigen synthesis occurs only in association with stem elongation. Thus, exogenous gibberellins promote stem elongation (see p. 46), and their stimulatory effects on flower initiation are indirect (i.e., gibberellins are required to bring about conditions needed for florigen synthesis). In SDP, and in those LDP whose stems elongate even under short days, gibberellins do not appear to be limiting for flower initiation, and we must presume that in these florigen is not synthesized under noninductive conditions for some other, unknown, reason.

### Effects of other growth hormones in flower initiation

It is not possible at present to ascribe any specific effect of the cytokinins, abscisic acid (ABA) or ethylene, upon floral induction, although their addition to plants may enhance flowering which has been induced by some other treatment (e.g., photoinduction or vernalization). Thus, kinetin, adenine, or zeatin treatment of SDP (e.g., *Perilla, Pharbitis, Wolffia microsopica*) increases the sensitivity of the plants to short days, but in long days SDP rarely flower in response to cytokinin treatment. It is known from other work that photoinduction requires nucleic acid synthesis, and it is possible that cases of cytokinins increasing flowering responses result from their general stimulatory effect on the production of nucleic acids (chapter 6). No evidence is available which suggests that endogenous cytokinins are concerned in the normal control of flower initiation.

Exposure of some plants to ethylene has revealed that this gas has effects on flowering similar to those elicited by auxins. Pineapple plants can be induced to flower by ethylene as well as by auxin treatment, and *Xanthium pensylvanicum* fails to form flowers even under short days in the presence of either auxin or ethylene. The similarity

---

so that the long-day partner was maintained under short days and the short-day partner under long days; both remained vegetative. (b) Transmission from induced short-day stock (*N. tabacum* cv. Maryland Mammoth) to a noninduced long-day plant (*Hyoscyamus niger*, annual variety) grafted on as scion. *Left.* Short-day partner on long days, long-day partner on short days; no flower formation. *Right.* Both under short days; short-day partner induced and noninduced long-day partner flowering (flowers on donor stock removed).
(a and b, A. K. Kudairi and A. Lang, unpublished work. Original prints supplied by Professor Anton Lang.)

of influences exerted by ethylene and auxins is referred to elsewhere (chapters 1, 2 and 5), together with the suggestion that a number of responses by plants to high auxin concentrations are consequences of increased ethylene evolution in the auxin-treated tissues. Thus, there seems no reason to believe that ethylene, any more than auxins, is directly concerned in the processes of flower initiation.

Abscisic acid has been reported to induce flowering in certain SDP (*Ribes nigrum*, *Pharbitis*, and *Fragaria*) when these are held under long days. However, ABA does not induce flowering when applied to many other typical SDP such as *Xanthium*, Biloxi soybean or Maryland Mammoth tobacco, and we must reserve judgment on the question of the possible role of ABA in the process of flower initiation in SDP. In LDP it has been found that ABA can inhibit flowering.

## Flower development

A shoot apex, once set on course to produce flower rather than vegetative organs, is the scene of precisely regulated events that lead to the appearance of a fully formed characteristic flower. Most angiosperms are monoecious, bearing hermaphrodite flowers. Development of bracts, sepals, petals, stamens, and finally carpels, takes place inexorably. The development of a flower is determinate; once the flower is fully formed the apical meristem ceases activity and often becomes incorporated into the structure of the carpels.

Very little is known of the controlling forces operative in a developing flower, just as we are still ignorant of the factors involved in the control of differentiation at a vegetative apex (p. 48). However, both auxins and gibberellins can, if applied to a flowering plant, modify the pattern of development in a flower. Treatment of hermaphrodite flowers of *Hyoscyamus niger* and *Silene pendula* with an auxin has been shown to suppress the development of both corolla and gynoecium (male parts, the stamens and anthers), resulting in flowers which are female only. Similarly, studies of dioecious species (male and female flowers on separate plants) such as hemp (*Cannabis sativa*), and of those monoecious species in which separate male and female flowers appear in different regions of the same plant (e.g., some varieties of cucumber, *Cucumis sativus*), have revealed that auxin or ethylene treatment cause the formation of female, as opposed to male, flowers. Conversely, treatment of gynoecious cucumber (i.e., a dioecious variety which normally produces only female flowers) with gibberellin results in the appearance of male as well as female flowers.

One may infer from the above observations that flower development occurs at least partially under the controlling influences of endogenous auxins and gibberellins. Indeed, some recent studies have revealed that in *Cucumis sativus*, maleness is associated with higher levels of endogenous gibberellins and lower levels of auxins than are seen in plants producing female flowers. In earlier studies by other workers, however, no correlation was found between endogenous auxin or gibberellin levels and flower sexuality.

## Fruit development

Except in those relatively rare plant species which have a natural tendency towards parthenocarpy (i.e., to produce sterile, seedless, fruits), fruit development occurs only after successful pollination. The arrival of pollen grains on the surface of the stigma is followed by growth of the pollen tube down through the style and into the ovary where the male nuclei are released. In flowering plants, fusion of one of the male nuclei with the egg cell forms the zygote which is destined to develop into the future embryo. The second male nucleus fuses with the two polar nuclei in the embryo sac, giving rise to the primary endosperm nucleus which divides repeatedly to form the endosperm tissue which surrounds, sustains, and protects the developing embryo (Fig. 4.1). The embryo, together with the endosperm and covering layers formed from the integuments of the ovule, comprise the seed. In conjunction with the development of the seed, the fruit develops from the tissues of the ovary (except in the case of so-called 'false fruits' such as the strawberry, where the bulk of the fruit is formed from a very enlarged receptacle which carries the achenes, the 'true fruits', on its surface). Thus, successful pollination sets in motion an ordered sequence of developmental events which culminate in the mature seed-containing fruit. For this reason, effective pollination is said to result in *fruit-set*. Failure of the pollination mechanism normally results in the formation of a separation layer (p. 124) at the base of the flower stalk, and abscission, or shedding, of the sterile reproductive structure.

In developing fruits there are very close physiological relationships between ovule, or seed, and fruit tissues, and growth hormones play significant and essential parts in the control of the complex pattern of development in fruits.

The first evidence that growth hormones are concerned in fruit development came from observations by Fitting in 1909, who found that when a water extract of orchid pollen was applied to an unfertilized orchid flower, then the petals withered and the ovaries started to swell, just as if the pollen itself had been placed on the stigma. In other words, union of male and female nuclei was not essential for fruit-set to occur. It was later discovered that pollen extracts contain auxin, and that initial fruit-set is brought about by the provision of auxin by pollen to the ovular structures. Consequently, treatment of unfertilized flowers of many species with a synthetic auxin may induce fruit-set. Not all plants respond in this way, however, for quite a number of species fail to set fruit when unfertilized even if exogenous auxin is supplied. In some cases a gibberellin may successfully induce fruit-set whereas auxins fail to. There is some evidence that ethylene may also be involved in fruit-set, for exposure of orchid flowers to this gas causes petal withering and swelling of the ovary similar to that induced by auxins. It is possible that auxin coming from pollen grains induces ethylene synthesis in the style and/or ovular tissues, and that it is ethylene rather than auxin which triggers off the typical fruit-set responses (see also p. 42).

Although it seems that the initial swelling of the ovary and ovule characteristic of fruit-set is occasioned by the arrival of auxin from pollen, the further development of the fruit normally follows only if the male and female nuclei unite (with the exception of the relatively few naturally parthenocarpic species). Presumably, the quantity of auxin (and perhaps gibberellin, also, in some species) supplied by pollen is sufficient only for the earliest phase of fruit development seen at fruit-set. However, subsequent to fertilization, a developing fruit contains high concentrations of auxins, gibberellins, and cytokinins, which appear to originate in the developing embryo and endosperm. The final size of many fruits such as apples, pears, blackcurrants, strawberries and tomatoes can readily be seen to be correlated with the number of fertile seeds contained (Fig. 4.5). It is generally considered that the close relationship between seed and fruit development largely results from the production of growth hormones within the seed. Great difficulty has, nevertheless, been experienced in attempts to elucidate the respective parts played by

*Figure 4.5.* (a)–(d) Correlation between development of fertilized achenes and adjacent receptacle tissue in strawberry. (a) All except one fertilized achene removed. (b) Three fertilized achenes present. (c) *Left*, all achenes on; *right*, three vertical rows of achenes. (d) *Left*, two horizontal rows of achenes; *right*, all achenes on. (e) Auxin-induced parthenocarpy: *left*, control fertilized fruit; *centre*, all achenes removed; *right*, all achenes removed and receptacle treated with the auxin, $\beta$-naphthoxyacetic acid. (From J. P. Nitsch, *Amer. J. Bot.* **37**, 211–215, 1950. Original prints supplied by Professor J. P. Nitsch.)

the different growth hormones during seed and fruit development, and no generalized scheme has yet been devised to accommodate all the observations made on the fruits of a range of species.

The early period of fruit development is associated with a rapid rate of cell-division activity, particularly in the endosperm and embryo (in most fruits cell divisions cease in ovary and receptacle tissues at, or shortly after, anthesis). The major increase in fruit volume is brought about by the vacuolation (enlargement) of the cells formed during the earlier phase of mitotic activity.

There have been two main ways of approaching the problem of the roles of growth hormones in the control of fruit development; (i) to study the effects on fruit development of removing young seeds or achenes (or to prevent pollination which achieves the same result), and attempting to reproduce seed influences with exogenous hormones, and (ii) to correlate stages of fruit and seed development with changes in levels of endogenous hormones.

Preventing pollination, or removing all seeds from a fertilized fruit, brings development to a halt. In some cases (e.g., in cucurbits and strawberry) auxin can substitute for seeds and allow the development of a parthenocarpic fruit (Fig. 4.5), but in others auxin is unable to elicit this response whereas gibberellin can (e.g., *Prunus* species such as almonds, cherries and peaches, and in grape, *Vitis vinifera*; Fig. 4.6). In yet other examples, parthenocarpic fruits form readily only when *both* an auxin and a gibberellin are applied (e.g., in unpollinated 'Bing' cherry, $GA_3$ induced fruit-set, but the fruits failed to develop fully and died. Addition of an auxin, which on its own was unable to cause even fruit-set, along with the gibberellin resulted in the formation of virtually normal cherries). All these observations suggest that development of fruit tissues, whether they be ovary or receptacle in origin, occurs in response to the transmission of auxin and gibberellin from developing seeds.

Studies of the endogenous hormones of fruits have shown that during the early period of development when cell-division rates are high, there are much greater concentrations of cytokinins present than later (e.g., in young fruits of apple, plum and tomato). The cell-enlargement phase of fruit development is characterized by high concentrations of endogenous auxins and gibberellins, and these can sometimes, but not always, be seen to correlate with the rate of fruit growth (Fig. 4.7). These findings support the general conclusion reached from studies of parthenocarpic fruit development in response to auxin and gibberelin applications, that both auxins and gibberellins are concerned in the induction of the cell-enlargement phase of fruit development. The auxins involved appear to be synthesized in the developing seeds, but although the same may usually be true for the gibberellins also, there is the possibility that some of the gibberellins present in expanding fruit tissues come from vegetative regions of the plant. This is suggested by the discovery that naturally parthenocarpic (seedless) grapes contain gibberellins.

*Figure 4.6.* Induction of parthenocarpic apple fruits (cv. Wealthy) by gibberellins. Left to right: normal seed-containing fertilized fruit; parthenocarpic fruit resulting from treatment with $10^{-2}$ M $GA_4$ in lanolin; parthenocarpic fruit induced by $10^{-2}$ M $GA_3$ in lanolin. Gibberellin $A_4$ is approximately five times more active than $GA_3$ in inducing parthenocarpy in this apple variety.
(From S. H. Wittwer and M. J. Bukovac, *Proc. Campbell Soup Co. Plant Science Symp.*, 1962, pp. 65–85. Original print supplied by authors, Michigan State University.)

During the last part of cell enlargement in fruits there is considerable accumulation of sugars which are obtained from the leaves. At this time it may be that the availability of sugars, rather than growth hormones, regulates the rate of fruit growth. Thus, it is possible that the increased osmotic pressure of fruit tissue cells resulting from the inflow of sugars causes osmotic water uptake and stretching of the cell walls. Nevertheless, even if this is the case we know from other evidence (see p. 145) that auxin is required for the cell wall softening which is essential for an increase in protoplast volume during cell vacuolation.

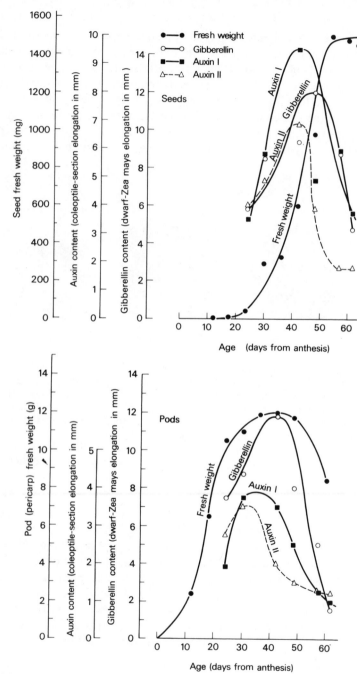

The final acts in fruit development are ripening and senescence, but this is dealt with separately in chapter 5. The ripe fruit is shed from the plant (though fruit shedding may also take place at earlier stages in its development) by an abscission process, and this subject, too, is also considered in chapter 5.

## Further reading list

1. Evans, L. T. (ed.) *Induction of Flowering*, MacMillan, New York and London, 1969.
2. Hillman, W. S. 'Photoperiodism and Vernalization', in *The Physiology of Plant Growth and Development* (ed. M. B. Wilkins), pp. 557–601, McGraw-Hill, London, 1969.
3. Lang, A. 'The physiology of flower initiation', *Encyc. Plant Physiol.* **15** (1), 1380, Springer-Verlag, Berlin, 1965.
4. Luckwill, L. C. 'Hormonal Aspects of Fruit Development in Higher Plants', *Symposia Soc. Exper. Biol.* **11**, 63, 1967.
5. Nitsch, J. P. 'Plant Hormones in the Development of Fruits', *Quart. Rev. Biol.* **27**, 33, 1952.
6. Salisbury, F. 'Photoperiodism and the flowering process', *Ann. Rev. Plant Physiol.* **12**, 293, 1968.
7. Wareing, P. F. and I. D. J. Phillips. Chapters 6, 9 and 10 in *The Control of Growth and Differentiation in Plants*, Pergamon Press, Oxford, 1970.

*Figure 4.7.* An example of the correlations which have been recorded between seed and fruit growth on the one hand, and the concentrations of endogenous auxins and gibberellins present in the tissues. Two unidentified auxins (Auxin I and Auxin II) and one gibberellin were detected in developing bean (*Phaseolus vulgaris*) seeds and pods, and the levels of these followed after anthesis and fertilization.
(Adapted from P. F. Wareing and A. K. Seth, *Symp. Soc. Exper. Biol.* **21**, 543–558, 1967.)

# 5. Growth hormones and phase change in plants

The term *phase change* may be applied to certain rather rapid transitions which occur in the normal pattern of plant development. The most marked phase change in plants is seen in the transition from the vegetative to the flowering condition (chapter 4). In some ways this dramatic alteration in developmental behavior is analogous to metamorphosis in animals. As we saw in the preceding chapter, known growth hormones are apparently not the principal factors which determine the vegetative-flowering phase change. However, a number of other phase changes occur during the life of a plant, the control of which more obviously involves the activities of endogenous growth hormones. These are *dormancy* in buds and seeds, *germination* of seeds, *senescence* and *abscission* of organs, and senescence of the whole plant.

## Dormancy and germination

Buds and seeds of many species exhibit dormancy. This is shown, in such species, by the fact that newly formed apical buds will not sprout, and freshly shed seeds will not germinate, even when placed under environmental conditions which are optimal for vegetative growth (i.e., adequate moisture, warmth and light).

The biological advantage of dormancy is that it allows plants to survive periods of unfavorable climatic conditions (i.e., cold winters in temperate zones, or hot dry summers in arid and some tropical areas). The dormant phase is often, but not always, characterized by a particular morphology, such as bulbs, corms, rhizomes, tubers and seeds. In many woody plants, dormancy is associated with the formation of resting-buds, the morphology of which differs from species to species, but typically consists of an apical meristem, leaf primordia, and un-elongated internodes, all surrounded by bud scales. Undoubtedly, these various morphological forms all afford the plant protection against a hostile environment, but resistance to frost and drought involves more than the possession of bud scales or retreat below the surface of the ground in the form of bulbs, corms, rhizomes and tubers. Distinctive biochemical changes occur during the onset of dormancy in a tissue, and these are reversed when the organ emerges from dormancy. In many instances a specific dormancy-breaking requirement must be met before growth resumes. This may take the form of a period of 'after-ripening' in dry storage of many seeds, but very commonly seeds and buds need to be subjected to several weeks of low temperature (0–2°C) under moist conditions ('chilling') for breaking of dormancy.

The onset of dormancy in many plants is hastened by exposure to short days (i.e., photoperiods of 10 hours or less in every 24 hour cycle), and delayed or prevented altogether by long days (approximately 16 hours or more light every 24 hours). Thus, the development of dormant buds in temperate-zone woody species is primarily a response to the natural short days in the fall. Some woody species (e.g., birch, *Betula pubescens*) will go on growing indefinitely if maintained under artificial long days, but others (e.g., English sycamore, *Acer pseudoplatanus*) cease growth eventually even under long days. Buds of birch may be released from dormancy by either a period of chilling or by exposure to long days.

Growth is suspended in dormant tissues because of the existence of a 'metabolic block', or blocks. Numerous normal metabolic reactions can be shown to be absent or suppressed during dormancy, but great interest centers upon identifying the 'key' which regulates the onset of, and emergence from, the dormant state. A better understanding of the factors involved in the control of dormancy will lead to an improved appreciation of the way in which plant growth generally is regulated. Considerable evidence has now accumulated which demonstrates that dormancy, as well as other aspects of growth and differentiation, is controlled by a mechanism involving changes in internal growth hormone levels.

**Bud dormancy**

Following the discovery of auxin, it was proposed that dormant tissues were unable to grow because of a deficiency in auxin. However, no consistent correlation has been found between auxin concentration in buds or seeds and their state of dormancy. Also, treatment of dormant buds or seeds with auxins does not normally induce them to grow; indeed, treatment of buds with an auxin may prolong rather than shorten the dormant period. The discovery by Hemberg, in 1949, that dormant terminal buds of ash trees (*Fraxinus excelsior*) and dormant potato (*Solanum tuberosum*) tuber buds contain higher levels of growth-inhibiting substances than similar buds emerging from dormancy, led to the concept that dormancy in plants is controlled by the presence of *growth inhibitors*. Considerable research effort during the 1950's and '60's by workers in various parts of the world has indicated that this proposal is essentially correct (Fig. 5.1). In the case of birch and sycamore for example, it was found that the level of growth inhibitor in mature leaves and shoot apices was always higher under short days which induced dormancy, than under long days when growth was maintained, or dormancy broken (Fig. 5.2).

The growth inhibitor present in sycamore was isolated and chemically characterized by Wareing, Cornforth and associated workers. It was found to be abscisic acid (Fig. 1.23), and Wareing has demonstrated that ABA not only inhibits growth in various bioassays (Fig. 1.22), but it will also induce dormancy in woody species. Thus, birch, sycamore and blackcurrant (*Ribes nigrum*) plants all ceased internode extension growth and formed typical resting-buds under long days (which favor continued growth) when ABA was added to the leaves or stem. In other words, ABA simulated the effect of short days.

There appears, therefore, good reason to consider that bud dormancy in many species is controlled through changes in levels of endogenous growth inhibitors, such as ABA, and that daylength and chilling affect dormancy by influencing growth-inhibitor biosynthesis or inactivation. More work is required before we can be sure that measured changes in endogenous growth-inhibitor levels, in response to photoperiod, are really reflections of effects on ABA biosynthetic or inactivation mechanisms.

In addition to growth inhibitors, such as ABA, other growth hormones, particularly the gibberellins, may also be involved in the

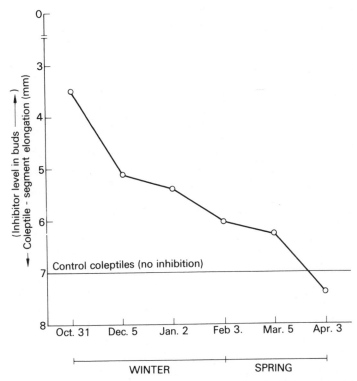

*Figure 5.1.* Decrease in endogenous growth-inhibitor (perhaps abscisic acid) buds of English sycamore (*Acer pseudoplatanus*) during the course of the winter and spring. At each date 0.1 g dry weight of bud tissues was extracted. Each extract was chromatographed, and the inhibitory fraction assayed in the wheat coleoptile-section straight growth test. The graph is reversed, in that lower levels of coleoptile elongation in the presence of higher concentrations of inhibitor are shown further up the axis. From early February onwards the buds were non-dormant, and commenced growth.

(From I. D. J. Phillips and P. F. Wareing, *J. Exper. Bot.* **9**, 350–364, 1958.)

control of bud dormancy. Treatment of dormant buds of a number of species with gibberellic acid (GA₃) results in their immediate growth (Fig. 5.3). Also, endogenous gibberellin levels have been found to rise at the time of their emergence from dormancy (Fig. 5.4). Since growth inhibitor levels appear to fall during emergence of buds from dormancy, it is possible that some sort of antagonistic

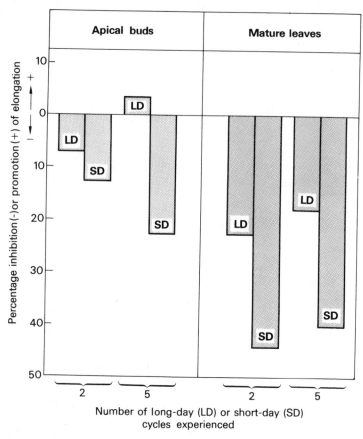

*Figure 5.2*. Higher levels of growth inhibitor can be detected in leaves and buds of woody plants maintained under short days, than in those kept under long days. Extracts of *Acer pseudoplatanus* buds and leaves were prepared as described in Fig. 5.1, except that 0.17 g dry weight of buds, and 1.0 g of mature leaves were used.
(From I. D. J. Phillips and P. F. Wareing, *J. Exper. Bot.* **10**, 504–514, 1959.)

interaction occurs between endogenous inhibitors and gibberellins; the latter hormones tending to induce growth, and inhibitors, such as ABA, to suppress growth and induce dormancy. Lending support to such a concept is the observation that the dormancy-breaking effect of $GA_3$ on birch bud dormancy is canceled if ABA is added at the same time.

All these observations agree with the view that bud dormancy is induced under conditions of low internal gibberellin levels (e.g., under short days, p. 95) and high growth inhibitor (perhaps ABA) levels, and that release from dormancy results from the reverse situation. Even so, we need still further information to reach any conclusion as to the validity of the gibberellin/inhibitor hypothesis for the control of bud dormancy.

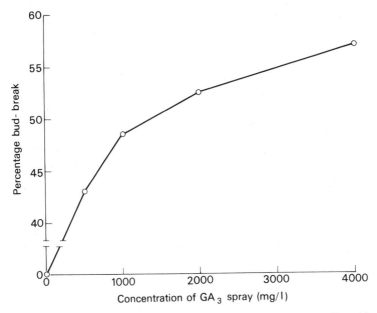

*Figure 5.3*. Effect of sprays of various concentrations of gibberellic acid (in 50 per cent ethanol) on bud growth in dormant (unchilled) 'Pullar' peach twigs. Sprouting of buds recorded 7 days after treatment. (From I. D. J. Phillips, *J. Exper. Bot.* **13**, 213–226, 1962.)

**Seed dormancy**

Research into the problem of seed dormancy has led to views of a hormonal control mechanism similar to that envisaged for bud dormancy. Many dormant seeds have been found to contain growth and germination inhibitors, and in some species it has recently been discovered that ABA is the principal inhibitor present.

Those seeds which normally require light for germination (e.g., birch, and some varieties of lettuce) may be caused to germinate in

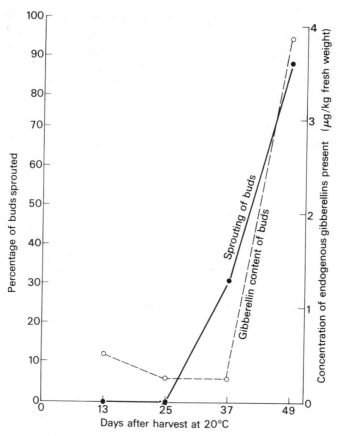

*Figure 5.4*. Relationships between gibberellins in, and sprouting of, potato tuber buds during after-ripening.
(From L. Rappaport and O. E. Smith, in *Eigenschaften und Wirkungen der Gibberellins*, ed. Rüdiger Knapp, 37–45. Springer-Verlag, Berlin, 1962.)

darkness by addition of a gibberellin, and ABA at least partially prevents this effect of gibberellins (Fig. 5.5). Similarly, seeds which in nature need to be chilled (or 'stratified') at 0–5°C for a period of time to break their dormancy, can also be induced to germinate by gibberellin treatment, and here again ABA reduces gibberellin promotion.

Evidence that growth- or germination-inhibitor levels decrease in seeds during their emergence from dormancy has been obtained for a number of species. Less conclusive evidence is available for a rise in gibberellin levels at the same time. In fact in hazel seed it was found by J. W. Bradbeer that endogenous gibberellin levels did not rise appreciably until *after* the requisite period of chilling was given to break dormancy, and the seeds were starting to germinate at a higher temperature. This suggests that the rise in endogenous gibberellin of the seeds was a result, rather than cause, of release from dormancy. Similar findings have been recorded for endogenous gibberellin levels in buds (see Fig. 5.4), and these, together with Bradbeer's results, cast some doubt on the validity of the gibberellin/ inhibitor hypothesis for dormancy control.

In conclusion, it is probably naïve to consider that dormancy in buds and seeds is controlled solely through a 'balance', or inter-action, of endogenous growth inhibitors, such as ABA, and gibberel-lins. Other features of dormancy which cause one to be cautious of the simple gibberellin/inhibitor hypothesis, include the fact that a number of substances other than gibberellins can, in some cases, break seed dormancy. Among these are the cytokinins (and ABA antagonizes cytokinin-stimulated germination as effectively as gibberellin-stimulated germination), but we have no good evidence yet which suggests that endogenous cytokinin levels in seeds are correlated with depth of dormancy.

**Germination**

Given adequate moisture and temperature, germination takes place in nondormant seeds. Germination of a seed, and subsequent early seedling growth, involves the consumption of food reserves which were deposited in the endosperm and/or cotyledons during seed development in the previous season. The mobilization of these food reserves to the growing embryo is, in at least some species, apparently initiated by growth hormones produced in the embryo.

The relationship between embryo and stored food at germination has been studied most intensively for barley (*Hordeum*) grains, where it is found that if the embryo is removed from the seed, starch reserves in the endosperm cells are not changed when the embryo-less seed is placed in moist surroundings (Fig. 5.6). Normally, in the presence of the embryo, endosperm starch is hydrolysed to reducing sugars which are translocated into the developing embryo. Starch digestion in barley seed endosperm is a result of the activity of the amylase enzymes. $\beta$-amylase is present in ungerminated barley seed,

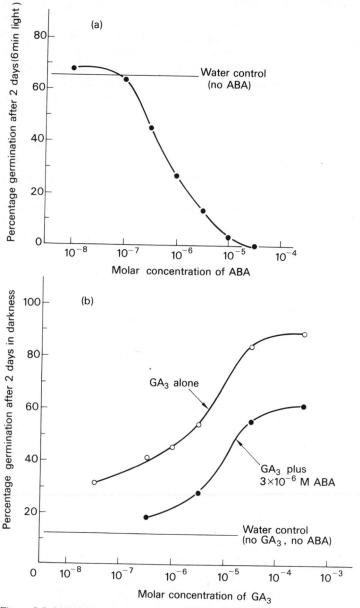

*Figure 5.5.* (a) Inhibitory effect of abscisic acid (ABA) on light-stimulated germination (6 min red light after 2 hr imbibition) of 'Attractie' lettuce seed. (b) Effect of ABA on gibberellic acid-stimulated germination in darkness of the same variety.

(Unpublished data of J. W. Bradbeer, King's College, London University.)

*Figure 5.6.* Enhancement by gibberellic acid ($GA_3$) of digestion of barley endosperm inwards from the aleurone cells, resulting principally from the action of α-amylase. From bottom to top: treated with 5 μl water only; with 5 μl $10^{-9}$ M $GA_3$; with 5 μl $10^{-7}$ M $GA_3$.
(Original print supplied by Professor J. E. Varner, Michigan State University.)

but α-amylase appears only at germination. In addition to α-amylase, other enzymes are active in the endosperm at germination. These include proteolytic enzymes and nucleases, which break down the proteins and nucleic acids, respectively, of the endosperm cells. Most of these enzymes are synthesized in the aleurone cells which surround the endosperm, and are secreted into the latter tissue. The actual synthesis of enzymes in the aleurone cells starts with the arrival of gibberellins coming from the germinating embryo (Fig. 5.7). Hence, the fact that excision of the embryo prevents endosperm digestion is due to the removal of the source of endogenous gibberellin. This is clearly demonstrated by treating embryo-less barley or wheat (*Triticum*) seeds with gibberellic acid, which induces the synthesis of α-amylase and other enzymes in the aleurone cells (Figs. 5.6 and 5.7). However, in wheat seeds the endosperm itself appears to play a part in the induction of hydrolytic enzymes in the aleurone cells, for in this species $GA_3$ fails to induce α-amylase synthesis in aleurone separated from endosperm, unless a cytokinin is supplied prior to the $GA_3$. Perhaps, therefore, cytokinins are formed in the imbibed endosperm and act on the aleurone, predisposing the cells of the latter tissue to react to gibberellins arriving from the embryo (Fig. 5.7).

## Senescence

Prior to death in a multicellular plant, the process known as senescence occurs. In senescing cells, there is a gradual reduction in the capacity for protein synthesis together with degeneration of the endoplasmic reticulum, ribosomes, mitochondria and other organelles and membranes. The rate at which these changes occur varies greatly, depending on the species, the part of the plant involved, and environmental conditions. Senescence may take place at more or less the same time in all parts of a plant, or individual organs may senesce whilst the remainder of the organism retains vitality. The end of senescence is marked by death.

---

*Figure 5.7.* Involvement of gibberellin from the embryo, and perhaps also cytokinin from the endosperm, in the induction of hydrolytic enzyme synthesis in cereal seed aleurone cells. I–IV. Possible sequence of events in imbibed seeds of wheat and barley. *Below.* Synthesis of α-amylase in wheat aleurone tissue. *Left.* Addition of kinetin increased the response of aleurone cells to $GA_3$ after a short period of imbibition. *Right.* After a longer period of imbibition, it appears that the aleurone tissue is self-sufficient in cytokinin, perhaps derived from the endosperm.
(Data in graphs from D. Eastwood, R. J. A. Tavener, and D. L. Laidman, *Nature* **221**, 1267, 1969.)

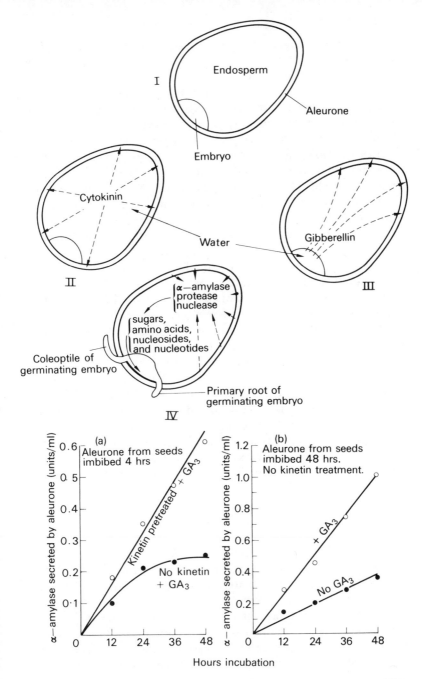

I

Endosperm

Aleurone

Embryo

II

Cytokinin

Water

Gibberellin

III

IV

α—amylase
protease
nuclease

sugars,
amino acids,
nucleosides,
and nucleotides

Coleoptile of
germinating embryo

Primary root of
germinating embryo

(a)
Aleurone from seeds
imbided 4 hrs

Kinetin pretreated + GA₃

No kinetin
+ GA₃

(b)
Aleurone from seeds
imbibed 48 hrs.
No kinetin treatment.

+ GA₃

No GA₃

α—amylase secreted by aleurone (units/ml)

Hours incubation

117

*Whole-plant senescence* is normally seen only in monocarpic plants (i.e., those which flower and fruit only once in their life cycle and then die), including all annuals and biennials and a few perennials such as the 'century plant' (*Agave*) and the bamboos (*Bambusa*). The death of long-lived woody perennials which flower repeatedly (i.e., polycarpic species) eventually occurs following the slow process of ageing. Ageing in trees may take many years, and is revealed by a successively lower annual growth increment and lessened resistance to pathogenic invasion. This is in contrast to senescence, which is a relatively rapid process, whether we are considering whole-plant or organ senescence. Thus, for example, senescence of annuals occurs quite quickly following fruiting, and leaf senescence in deciduous trees takes place within a few weeks in the fall. *Organ senescence* occurs in both monocarpic and polycarpic species. Leaf senescence and fruit senescence are good examples of organ senescence.

In almost all plants each leaf has only a limited lifespan. As the shoot extends upwards in monocarpic plants, older leaves at the more basal end of the stem become senescent and die. A 'wave' of leaf senescence therefore passes up the stem, and this pattern is known as *sequential senescence*. In contrast, leaf senescence in polycarpic plants typically involves all the leaves at the same time (e.g., the dropping of leaves in the fall), and is consequently termed *synchronous*, or *simultaneous senescence*. Senescence of fruits is a late stage in the ripening process, and does not commence until the developing seeds are fully formed.

Clearly, senescence is a distinct physiological and biochemical phase. Its timing is controlled by both internal and external factors. Whole-plant senescence typically occurs when fruiting has finished, but in polycarpic plants leaf senescence, for example, occurs more in response to short days and lower temperatures than to internal factors associated with reproduction.

## Leaf senescence

Despite the diversity in patterns of senescence, it is likely that growth hormones are intimately concerned with the occurrence of both whole-plant and organ senescence. Much of the evidence in support of this view has come from studies of senescence in excised leaves and disks of lamina tissue. Separation of a leaf, or a leaf disk, from the parent plant normally results in the immediate onset of senescence in the excised tissues. This may be measured in various ways, but the decrease in protein and chlorophyll levels is usually used as an estimate of the progress of senescence (Fig. 5.8). Ultimately, all chlorophyll is lost and the detached leaf tissues become yellow in color. Addition of one of the growth-promoting hormones may delay or completely prevent senescence in detached leaves. A cytokinin, such as kinetin, effectively stops senescence in detached

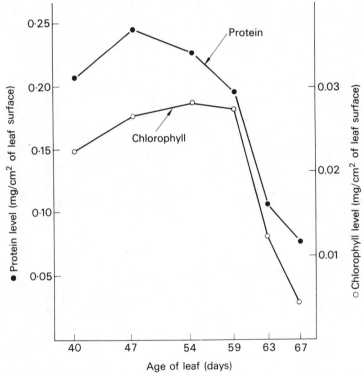

*Figure 5.8.* Changes in protein and chlorophyll levels with age in attached leaves of *Perilla*.
(From H. Woolhouse, *Symp. Soc. Exper. Biol.* **21**, 179–214, 1967.)

leaves of a great many species, so that the treated excised leaf tissue retains its green color and protein (Fig. 5.9). In leaves of certain other species, such as dock (*Rumex*), dandelion (*Taraxacum*) and nasturtium (*Tropaeolum*), senescence can be delayed at least as effectively by exogenous gibberellins as by exogenous cytokinins (Fig. 5.9). Auxins are generally ineffectual in delaying senescence in detached

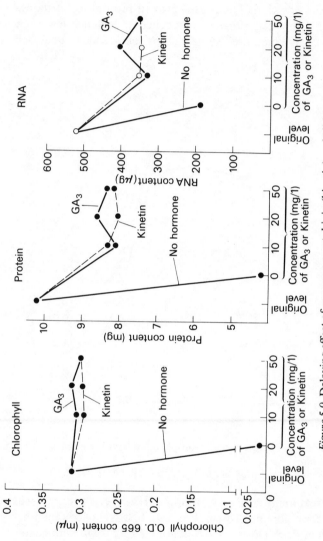

Figure 5.9. Delaying effect of exogenous cytokinin (kinetin) and gibberellin (GA$_3$) on senescence in excised disks of nasturtium (*Trapaeolum majus*) leaves. Ten leaf disks were used in each treatment, and chlorophyll, protein and RNA contents measured immediately on excision, and after 10 days.

(From data of L. Beevers and F. S. Guernsey, *Nature* **214**, 941–942, 1967.)

leaves of herbaceous species, but the senescence of both detached and attached leaves of deciduous trees can be delayed by auxin treatment (Fig. 5.10). The reasons for leaves of different species responding differently to the different categories of growth hormone are not yet clear.

There is evidence that roots produce a substance or substances essential for the maintenance of protein and chlorophyll levels in leaves. Thus, in 1939 Chibnall observed that senescence did not occur in detached leaves when roots were initiated in and grew from the petiole, and he postulated the existence of a 'root factor' which is required for continued protein and chlorophyll synthesis. The discovery by Richmond and Lang in 1957, that kinetin can substitute for roots in preventing senescence in detached *Xanthium* leaves, suggested that Chibnall's root factor was a cytokinin. We now know that cytokinins are indeed produced in roots and exported to the shoot, for their presence has been detected in the 'bleeding-sap' which exudes under root pressure from root systems severed from the shoot. Nevertheless, a number of workers have found that sequential senescence of *attached* older leaves cannot be prevented by kinetin treatment, which suggests that something in addition to cytokinin is concerned in the control of senescence in attached leaves. The reason for kinetin delaying senescence in detached but not in attached leaves is not clear. One factor involved, however, is that an attached leaf is subject to correlative influences exerted by younger regions of the shoot, and is gradually drained of metabolites by the younger parts. Thus, a synthetic cytokinin applied directly to an attached old leaf may be rapidly lost to younger leaves. In detached leaves, on the other hand, applied cytokinin would be retained. One recent report, however, showed that application of 6-benzylaminopurine (BAP, a synthetic cytokinin) to primary leaves of intact bean plants did delay senescence of both the leaves and entire shoot.

The contrast between the ineffectiveness of auxins in delaying senescence in leaves of herbaceous plants, and their ability to prevent senescence in tree leaves, suggests that sequential and synchronous leaf senescence are differently controlled. This is perhaps not surprising, in view of the fact that sequential leaf senescence is apparently determined by internal correlative factors such as competition for nutrients and cytokinins, whereas synchronous leaf senescence is triggered by environmental factors (p. 118).

In contrast with the senescence delaying effects of auxins, cytokinins and gibberellins, senescence of isolated leaf disks is promoted by ABA (Fig. 5.10). It is, however, not known whether natural leaf senescence is in any way determined by endogenous ABA. Another plant hormone, ethylene, also promotes leaf and fruit senescence, but here again we are not at all clear yet whether, or how, endogenous ethylene is involved in the determination of normal patterns of senescence.

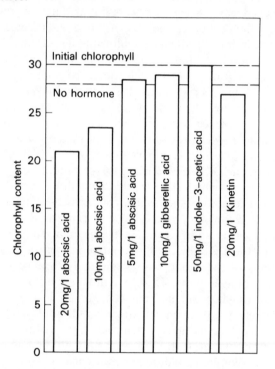

*Figure 5.10.* Acceleration of senescence in excised leaf disks of *Acer pseudoplatanus* by abscisic acid, and its retardation by indole-3-acetic acid, over a period of 10 days.
(From H. M. M. El-Antably, P. F. Wareing, and J. Hillman, *Planta* **73**, 74–90, 1967.)

**Fruit senescence**

Up until the time of seed maturity, fertile developing fruits show no signs of senescence, but shortly after this time the ripening process starts, which culminates in senescence and eventual death and decay of the fruit tissues (Fig. 5.11). If, however, the developing seeds are removed from young fruits, then the pericarp tissues senesce. Supplying an auxin to de-seeded fruits prevents their premature senescence, and it appears that this is another example of auxin-directed transport (p. 63), in that high auxin levels in developing fruits enable these structures to obtain a supply of necessary metabolites, including, perhaps, a flow of cytokinins from the roots.

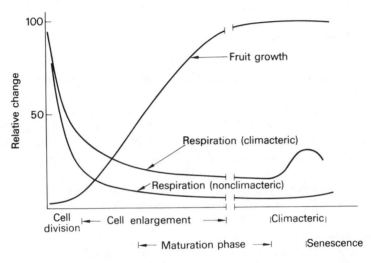

*Figure 5.11.* Stages in fruit development and maturation, and respiratory trends in climacteric and nonclimacteric fruits. The discontinuity in lines indicates the variable time scale. The growth curve may be single-sigmoid (as illustrated) or double-sigmoid.
(From J. B. Biale, *Science* **146**, 880–888, 1964. Copyright 1964 by the American Association for the Advancement of Science.)

### Whole-plant senescence

As we have already considered (p. 118), whole-plant senescence normally follows flowering and fruiting. To a certain extent this is explicable in terms of competition between vegetative and reproductive organs for nutrients and substances such as cytokinins, and invoking auxin-directed transport in a similar manner to that proposed earlier in connection with apical dominance (p. 63). Thus, developing fruits are active centers of auxin synthesis and may therefore monopolize available metabolites to the detriment of vegetative regions. However, this does not adequately explain why in dioecious species such as spinach, male whole-plant senescence is correlated with the production of male flowers. Also, removal of male flowers as they appear delays senescence in vegetative regions. Thus, competition for metabolites, including cytokinins, cannot be the complete basis for the observed correlation between whole-plant senescence and reproduction. Much more work is required to resolve this problem.

## Leaf abscission

Shoot growth is characterized by elongation of the main and lateral stems, and at the same time new leaves are formed which proceed to expand. This causes the older leaves, situated lower down, to be subjected to increasingly shaded conditions. Consequently, leaves may become less 'useful' to the plant as time goes on, due to suppression of their photosynthetic activities by lack of light. This is, perhaps, one reason for the existence in plants of mechanisms whereby older leaves are dispensed with. In some species, particularly monocotyledons, old leaves simply wither and remain attached to the stem, but many dicotyledonous species have a system of leaf shedding, known as abscission, which is regulated by means of endogenous growth hormones.

Leaf senescence and abscission are not confined to the fall and deciduous trees, but go on all through summer in many annuals and also evergreen trees. In the tropics, leaves drop and new ones grow all through the year, and in some tropical species old leaves are shed and new ones produced immediately after (aptly called 'leaf-exchanging species'). Leaf abscission usually involves the formation of a *separation layer* situated transversely across the base of the petiole immediately adjacent to the stem (Fig. 5.12). The 'inter-cellular cement' between the cells of the separation layer is dissolved by enzymatic activity, so that the strength of attachment of leaf to stem is severely weakened. The leaf falls from the plant by mechanical rupture in the separation layer.

Characteristically, abscission is the final act in the process of leaf senescence (p. 118). There are certain exceptions to this, but even in these senescence proceeds to a certain extent before abscission occurs. Any factor which hastens senescence, therefore, also induces leaf abscission. Conversely, young active green leaves contain factors

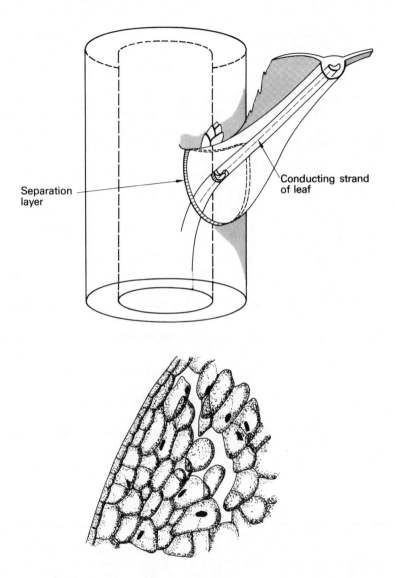

*Figure 5.12.* Separation layer formation at the base of the petiole during leaf abscission. *Above.* Position of the separation layer. *Below.* Section through the separation layer, showing cells coming apart during abscission.
(From L. J. Audus, *Plant Growth Substances*, 2nd ed. Leonard Hill (Books) Ltd., London, 1959.)

which prevent both their senescence and abscission. Great interest thus centers upon identifying senescence and abscission delaying factors in young leaves, and factors which speed up these processes in old leaves.

The formation of a separation layer is known to occur when the auxin-synthesizing capacity of the lamina drops to a low level (Fig. 2.9). As long as sufficient quantities of auxin from the lamina reach the lower end of the petiole, abscission is prevented. Removal of the lamina from the distal end of the petiole of a young leaf hastens abscission of the petiole. The reason for this is that removing the lamina also removes the principal source of auxin for the petiole. This is easily demonstrated in plants such as *Coleus* and cotton, where application of an auxin to the cut end of a debladed petiole prevents abscission (Fig. 5.13). More active auxin synthesis takes place in young than in old leaves (Fig. 2.9), which means that there is a correlation between low auxin level in the petiole and time of abscission.

Separation layer formation is not, however, simply a response to an absence of auxin in the leaf. Other hormonal factors are involved, for senescing leaves contain abscission-promoting substances. The growth-inhibiting hormone, abscisic acid (ABA), was first discovered by Addicott *et al.* as a factor which accelerated leaf abscission in cotton plants, even though it was actually extracted from immature cotton fruits (p. 37). Although ABA is effective in inducing cotton leaf abscission, even when as little as one microgram is applied, it does not seem to be particularly potent in producing leaf abscission in other species. Nevertheless, ABA levels have been reported to rise during senescence of nasturtium leaves, and we must consequently reserve judgment on the question of its possible involvement in the natural control of leaf abscission.

---

*Figure 5.13. Top.* A method commonly used for testing the effects of various substances upon leaf abscission. The substance under test may be applied to the cut ends of the petioles and/or to the stem. Species other than cotton may also be used (e.g., *Coleus*, see below). *Bottom.* Mean number of days taken for abscission of *Coleus* petioles, in relation to auxin (IAA) treatment. The auxin delayed abscission when applied to either the petioles (*distal*) or to the stem (*proximal*).
(Top, from F. T. Addicott, H. R. Carns, J. L. Lyon, O. E. Smith, and J. L. McMeans, *Régulateurs Naturels de la Croissance Végétale*, Centre Nat. de la Recherche Sci., Paris, 1964, pp. 687–703. Bottom, from W. P. Jacobs, *Plant Physiol.* **43**, (9B), 1480–1495, 1968.)

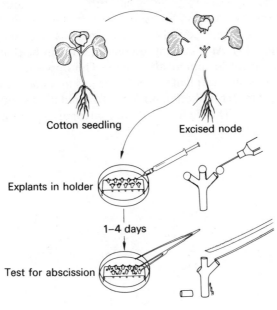

Cotton explant technique

Cotton seedling

Excised node

Explants in holder

1–4 days

Test for abscission

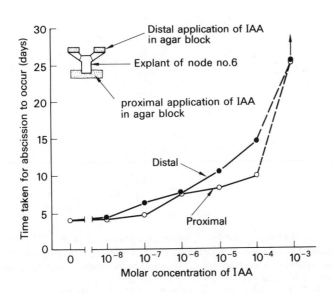

Distal application of IAA
in agar block

Explant of node no.6

proximal application of IAA
in agar block

Distal

Proximal

Time taken for abscission to occur (days)

Molar concentration of IAA

The natural abscission-accelerating factor in senescing leaves of many species appears to be ethylene. Ethylene is produced in all plant organs, but highest concentrations of the gas are found to emanate from young actively growing regions, including young leaves (Fig. 5.14). At first sight, therefore, it might be thought that ethylene is likely to inhibit abscission. The opposite is the case, however. Exposure of plants to ethylene gas at a concentration as low as one part per million in air leads to rapid abscission of older leaves (Fig. 5.15). The higher the auxin content of a leaf the less likely

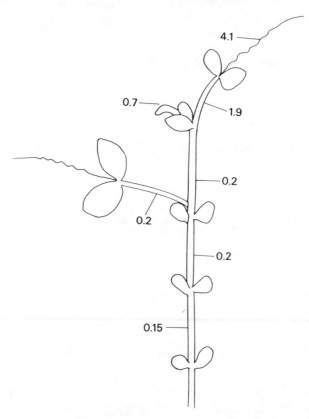

*Figure 5.14.* Ethylene production by various parts of a 14-day old green pea plant. The rates were determined for 5 mm-long pieces of stem and petiole tissues, or entire tendrils and apical hook, and are expressed as microlitres ethylene per g fresh weight of tissue per hour.
(From S. P. Burg, *Plant Physiol.* **43**, (9B), 1503–1511, 1968.)

is it to abscind ( = abscise) in the presence of ethylene. In the natural situation, young leaves contain high concentrations of both auxin and ethylene, whereas older leaves are low in both types of hormone. However, diffusion of ethylene from young leaves to old leaves may be expected to induce abscission of the latter. There is no doubt that the presence of young leaves *does* accelerate the abscission of old leaves, for excision of young leaves extends the life of older ones.

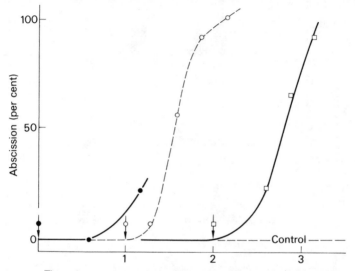

*Figure 5.15.* Effect of 0.25 ppm ethylene applied at various times (indicated by arrows) on abscission of cotton petioles.
(From S. P. Burg, *Plant Physiol.* **43**, (9B), 1503–1511, 1968.)

This may be a result of a reduction in ethylene concentration around old leaves consequent upon removal of young leaves, but other factors such as competition between leaves for nutrients, etc., are also likely to be involved. In fact, the acceleration of abscission of older leaves by the activities of young leaves may well be caused by the incapacity of auxin-deficient old leaves to 'attract' nutrients, which may flow to the auxin-rich young tissues (see p. 63).

In summary, both auxin and ethylene appear to be involved in the control of leaf abscission. Young auxin-rich leaves not only synthesize more ethylene, but they are also more tolerant of high levels of ethylene (natural or applied) than old leaves which are deficient in

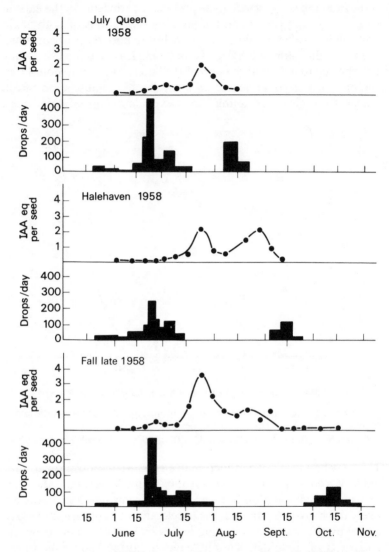

*Figure 5.16*. Correlation between endogenous auxin levels ('IAA equivalents') and incidence of abscission ('drops per day') in fruits of three peach varieties.
(From L. E. Powell and Charlott Pratt, *J. Hort. Sci.* **41**, 331–348, 1966.)

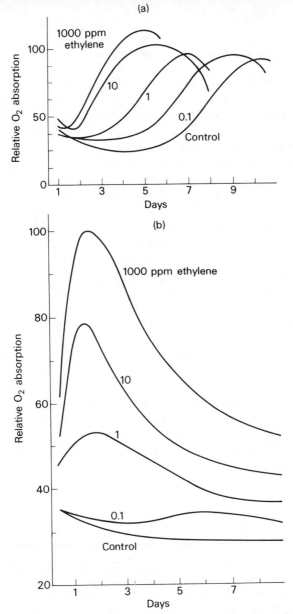

*Figure 5.17.* Oxygen uptake by developing fruits, and the effect of ethylene gas upon the normal respiratory trends. (a) For fruits which normally show the respiratory climacteric phenomenon (e.g., avacado pear). (b) For nonclimacteric fruits such as banana.

(From J. B. Biale, *Science* **146**, 880–888, 1964.)

auxin. The most recent work on leaf abscission has revealed that, in fact, greatly increased quantities of ethylene are synthesized in tissues adjacent to the separation layer when a particular stage of senescence is reached. It is suggested that this ethylene initiates the biochemical events which culminate in abscission of the leaf.

## Flower and fruit abscission

Flowers and fruits, as well as leaves, abscind. Failure of the pollination mechanism is often followed by the formation of a separation layer at the base of the pedicel. Abscission of fruits is most likely to occur at particular stages of their development, and is usually correlated with periods in which there is a low auxin content of the fruit (Fig. 5.16). In some commercial apple varieties, there are three periods of tendency for fruits to fall off the tree. These are; immediately after pollination (called 'post-blossom' drop), at the end of fruit growth ('June drop'), and during ripening ('pre-harvest drop'). The role of auxin in flower and fruit abscission therefore appears to be similar to its role in leaf abscission.

Ethylene stimulates flower and fruit, as well as leaf, senescence and abscission, and is probably concerned in these phenomena. Thus, developing fruits which normally experience a respiratory climacteric (see Fig. 5.11) show an earlier onset of the respiratory rise when treated with ethylene, and normally nonclimacteric fruits exhibit a respiratory climacteric in the presence of ethylene (Fig. 5.17). Also, exposure of fruits and flowers to ethylene can lead to accelerated abscission of these organs. Ethylene production by ripening fruits, or even by pollinated flowers, has been found to be sufficient in some cases to cause their own abscission.

## Further reading list

1. Carns, H. R. 'Abscission and its control', *Ann. Rev. Plant Physiol.* **17**, 295, 1966.
2. Jackson, M. B., and Daphne J. Osborne. 'Ethylene, the Natural Regulator of Leaf Abscission', *Nature* **225**, 1019–1022, 1970.
3. Various articles in 'Aspects of the Biology of Ageing', *Symp. Soc. Exper. Biol.* **21**, 1967.
4. Wareing, P. F. 'Germination and Dormancy', in *The Physiology of Plant Growth and Development* (ed. M. B. Wilkins), pp. 603–644, McGraw-Hill, London, 1969.
5. Wareing, P. F. and I. D. J. Phillips. Chapters 11 and 12 in *The Control of Growth and Differentiation in Plants*, Pergamon Press, Oxford, 1970.

# 6. The mechanism of action of plant growth hormones

Earlier chapters will have made it clear that many aspects of plant development are influenced by the ubiquitous activities of auxins, cytokinins, gibberellins, abscisic acid and ethylene. Any attempt to understand the processes of plant development must therefore take into consideration endogenous growth hormones. One may regard development as the basic biological problem demanding explanation by man, for orderly growth, differentiation, and reproduction are the hallmarks of living creatures. Because of the fundamental importance of growth hormones in the tightly ordered sequence of events which together constitute development in plants, and the tremendous interest which lies in understanding the forces at work in developing organisms, strenuous efforts are being made in the exciting quest to understand in molecular terms the way in which growth hormones exert their regulatory effects in plant development.

Three main avenues have been followed in approaching the problem of the mechanism of plant growth hormone action. One is the examination of numerous analogues of hormone molecules, in the hope of identifying what attributes a molecule must possess to have the properties of a plant hormone. Such studies of *hormone structure–activity relationships* will, it is hoped, allow us to know something of the type of molecule within the cell (the 'receptor site') with which a hormone molecule reacts, and thereby afford an insight into the manner in which hormones influence the cellular machinery.

The second approach to the problem is to examine the *physical changes in plant cells* associated with growth, and to relate these to effects of growth hormones on the same physical aspects. The cell wall, a characteristic feature of plant cells, has received particular attention in this connection. The reason for this is that an increase in volume of a plant cell can come about only if the physical properties of its wall are such as to allow it to stretch.

As growth hormones have such profound effects upon the metabolic processes of plant cells, a third way to gain access to knowledge of the mechanism of hormone action is to make detailed studies of the time course of *biochemical changes* induced by hormones. A particular difficulty in such studies, although not peculiar to them, is to distinguish between cause and effect; the very multiplicity of metabolic reactions induced or modified by hormones, and the rapidity with which they appear, has ensured that researchers have been kept busy in attempts to single out the 'master reaction' between hormone and receptor site from which all the other reactions follow, and lead ultimately to morphologically visible development. Because in recent years tremendous and exhilarating advances have been made in our understanding of the control of protein synthesis by specific ribonucleic acids (RNA), which in turn are controlled by coded genetic information in the deoxyribonucleic acid (DNA) of each living cell, attention is currently focused upon interactions between hormones and nucleic acid and protein synthesis. This interest is of course justified, for the basis of growth and differentiation lies in ordered sequential production of enzyme proteins.

A fourth route taken by researchers interested in hormone action involves a somewhat more direct approach to the identification of the hormone receptor site in plant cells. Various methods have been employed in this, most of which make use of isotopically labeled hormones. These can be supplied to plant tissues, and when they have been taken up attempts made to pinpoint their location within individual cells. Greater difficulties than may be at first envisaged are to be encountered in this. Sectioning treated cells and autoradiographing the sections has been done quite a number of times, but with only limited success in relating the hormone to any specific cell organelle or molecule (Fig. 6.1). Only very recently have associations been observed between applied radioactive hormone molecules and particular molecules in the cell, and these were made not by autoradiographic methods but by examination of extracts of hormone-treated cells.

134

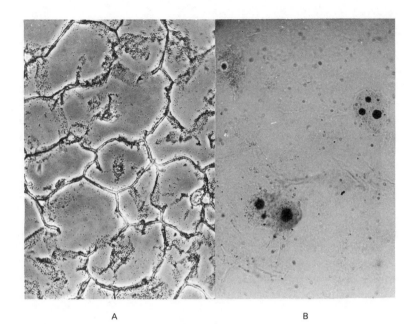

|     |     |
| :-: | :-: |
| A   | B   |

*Figure 6.1.* Distribution of radioactive growth hormones within cells from artichoke tubers. (A) Autoradiograph of tissue grown for 4 days in tritiated kinetin ($^3$H-kinetin). In some cells, intensity of labeling (blackened spots) was greatest over nuclei, but this was not consistently the case and no conclusion can be drawn as to the intracellular location of kinetin. (B) Autoradiograph of tissue grown for 4 days in a radioactive auxin (2,4-D-$^{14}$C). Almost all the nuclei in the tissue contained labeled nucleoli. Up to six nucleoli are present in each artichoke cell nucleus, and it was usually found that all of these nucleoli were labeled with $^{14}$C.

The localization of 2,4-D-$^{14}$C in nucleoli may have significance in relation to the known enhancing effect of this and other auxins on ribosomal-RNA synthesis, which takes place in nucleoli. On the other hand, exposure of artichoke tissues to 2,4-D-$^{14}$C for 6 hours, rather than 4 days, resulted in the labeling of only a few nucleoli, and yet 2,4-D enhances ribosomal-RNA synthesis within 6 hr.

(From J. A. Zwar and R. Brown, *Nature* **220**, 500–501, 1968. Original prints supplied by Dr. J. A. Zwar.)

## Structure–activity relationships of plant hormones

Studies aimed at identifying the chemical and physical features of hormonally active compounds have, in the main, been pursued because of the fundamental interest in understanding the mechanism of growth hormone action. In the case of the auxins, particularly, a high proportion of the total effort expended in this field has had very good practical and commercial motives, owing to the economic importance of certain synthetic auxins for use as selective weed-killers, root-inducing agents and fruit-set hormones. Much of the earlier work was of a largely empirical nature, involving the screening of a range of organic compounds to see whether or not any had the ability to produce auxin-like responses in higher plants. To a certain extent this screening approach is still used in the search for commercial auxins, but a fair body of knowledge has now been built up which allows 'educated guesses' to be made as to the types of molecules likely to have the properties of an auxin.

**Auxins**

Soon after the discovery that indole-3-acetic acid (IAA) is a native auxin in higher plants (p. 9) it was found that chemically related indole compounds such as indole-propionic, indole-butyric and indole-pyruvic acids (Figs. 1.3 and 6.2) all have biological activities

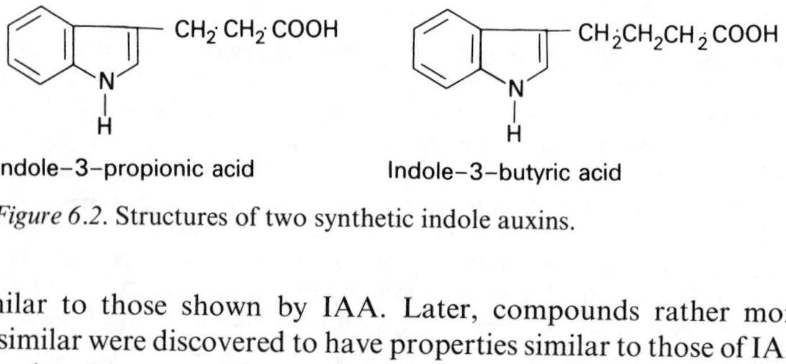

Indole–3–propionic acid          Indole–3–butyric acid

*Figure 6.2.* Structures of two synthetic indole auxins.

similar to those shown by IAA. Later, compounds rather more dissimilar were discovered to have properties similar to those of IAA in various bioassays. The most notable of these are certain phenoxy-acetic acid derivatives, such as 2,4-dichlorophenoxyacetic acid (2,4-D), 2,4,5-trichlorophenoxyacetic acid (2,4,5-T) and 4-chloro-2-methylphenoxyacetic acid (MCPA) (Fig. 6.3), all of which have proved of enormous value as selective herbicides (i.e., in relatively high concentrations they can kill many dicotyledonous weed plants without harming cereal or grass crops).

In the late 1930's, it was possible to describe in detail molecular requirements for auxin activity. At that time all known auxins were aromatic organic compounds (i.e., containing a ring system) with the ring system being unsaturated (i.e., at least one double bond in the ring), and with a side chain attached to the ring system. The side chain, it was considered, must terminate in an acidic carboxyl (—COOH) group, or be readily convertible into a carboxyl group, and that there must be at least one carbon atom between the carboxyl group and the ring system (i.e., the carboxyl group not to be attached directly to a carbon atom of the ring system). In addition, it appeared that an auxin molecule must have a configuration which allows a particular spatial relationship between the side chain and the ring system. These general 'structural requirements' for auxin activity hold true for analogues of indole acetic acid and phenoxyacetic acid (Figs. 1.3, 6.2 and 6.3). They do not hold, however, in the case of certain more recently discovered synthetic auxins, such as some benzoic acid derivatives and thiocarbamates (Fig. 6.4).

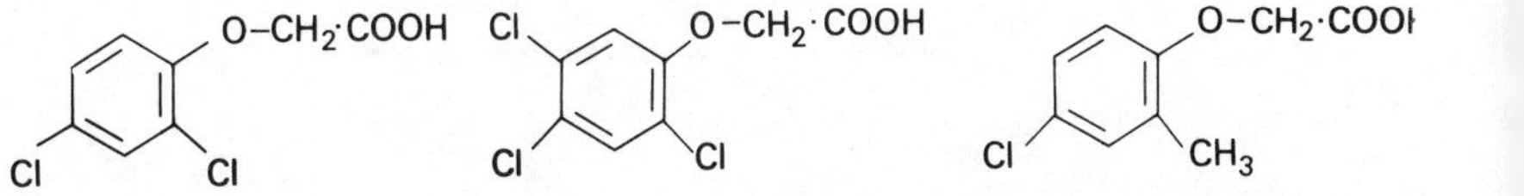

4-dichlorophenoxyacetic acid (2,4-D)

2,4,5-trichlorophenoxy-acetic acid (2,4,5-T)

2-methyl-4-chlorophenoxy-acetic acid (MCPA)

*Figure 6.3.* Three synthetic auxins widely used as selective weed-killers (herbicides). All are phenoxyacetic acid derivatives.

2,6-dichlorobenzoic acid

2,3,6-trichlorobenzoic acid

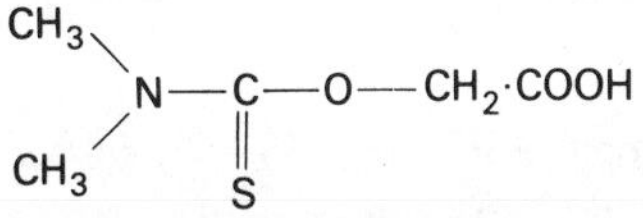

Carboxymethyl-thiocarbamate

*Figure 6.4. Above.* Two of the most active auxins in the series of benzoic acid derivatives. *Below.* One of the thiocarbamate compounds active as auxins.

Up until the mid-1950's, it was generally assumed that auxins become bound to a cellular receptor molecule, thought to be a protein, by chemical (covalent) bond formation. Studies of the structure–activity relationships had appeared to demonstrate that for activity as an auxin, a molecule had to comply with the list of structural features described above, and also have the ability to form covalent bonds at two, or more, specific positions on the molecule; (i) at the carboxyl end of the side chain, and (ii) at an *ortho* position on the ring system (i.e., on one of the two carbon atoms adjacent to the ring carbon connected with the side chain). This concept was known as the *two-point attachment theory* for auxin action and is illustrated for four phenoxy compounds in Figure 6.5. Despite the ingenuity of the

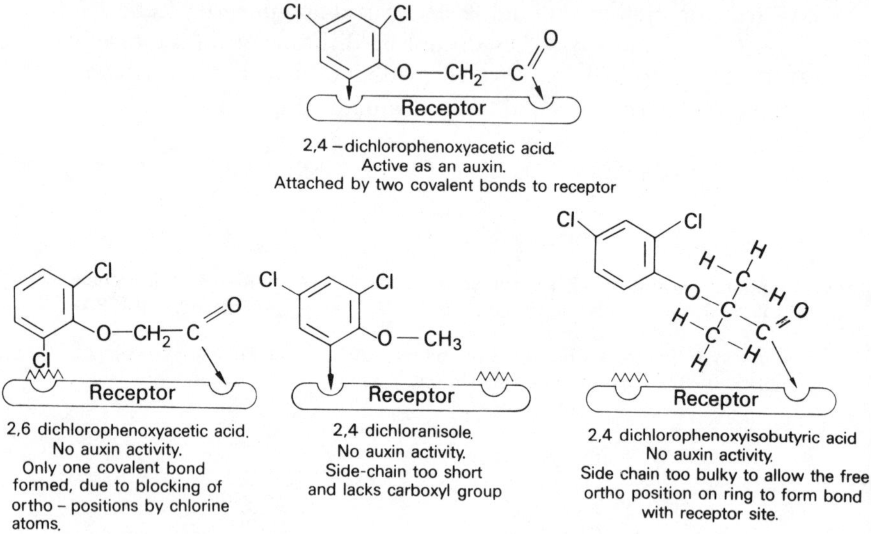

*Figure 6.5.* An illustration of the two-point attachment theory for auxin action, comparing 2,4-dichlorophenoxyacetic acid with three inactive analogues.

two-point attachment theory, and the rather good fit between it and observed structure–activity relationships at the time it was put forward, it now appears much more likely that auxins become associated with receptor molecules by purely physical means, rather than by covalent bonds. Electrostatic forces of attraction between receptor surface and auxin may be important. At present, therefore, the view is that it is the overall size and shape of a molecule which determines whether or not it acts as an auxin. An appropriate molecular configuration for auxin activity must be presumed to possess a pattern of charges which renders it compatible with the receptor site in the cell. We do not yet have a clear picture of the essential physical features of auxins, but a start has been made in that a comparison of a number of auxins of contrasting chemical structure has revealed that they have certain physical attributes in common (Fig. 6.6).

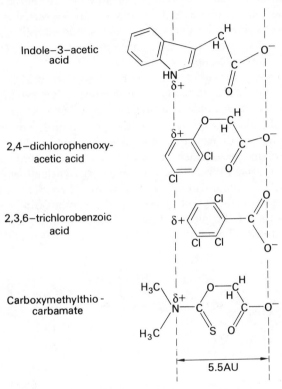

*Figure 6.6.* Similarity of four types of molecules which possess auxin activity, in that they each have a strong negative charge ( − ) situated 5.5 Angstrom units (5.5 AU) from a weaker positive charge ($\delta +$).
(From K. V. Thimann, *Ann. Rev. Plant Physiol.* **14**, 1–18, 1963.)

Thus, the prolonged and intensive attention devoted to structure–activity relationships of auxins has yielded a great deal of applied and commercial value, but has helped very little in unraveling the mechanism by which these substances can exert control over cellular metabolism.

## Gibberellins

All compounds known to be active as gibberellins in plants are based upon the gibbane carbon skeleton (Fig. 1.13). Some substances rather similar, but nevertheless different, in structure from the gibberellins (Fig. 6.7) do have slight biological activity in the manner of gibberellins, but careful study of the fate of some of these in plant tissues has shown that they can be enzymically converted to gibberellins. Thus, so far as we know at present, gibberellin-like biological activity is displayed only by compounds with the characteristic gibbane carbon skeleton.

Nevertheless, the twenty-eight gibberellins already identified from plants or fungi differ in the types of responses they evoke in higher plants. Thus, for example, in a single bioassay such as the elongation of cucumber hypocotyl, some of the gibberellins induce marked elongation growth whilst others have little or no effect. With a different plant species, such as a dwarf *Zea mays* mutant variety, for instance, a rather different picture can emerge, with gibberellins that are inactive in the cucumber test now revealing their capacity to promote elongation growth, and *vice versa*. Similarly, in a given plant species some of the gibberellins may elicit a flowering response whereas other gibberellins do not. Clearly, specificity of gibberellins

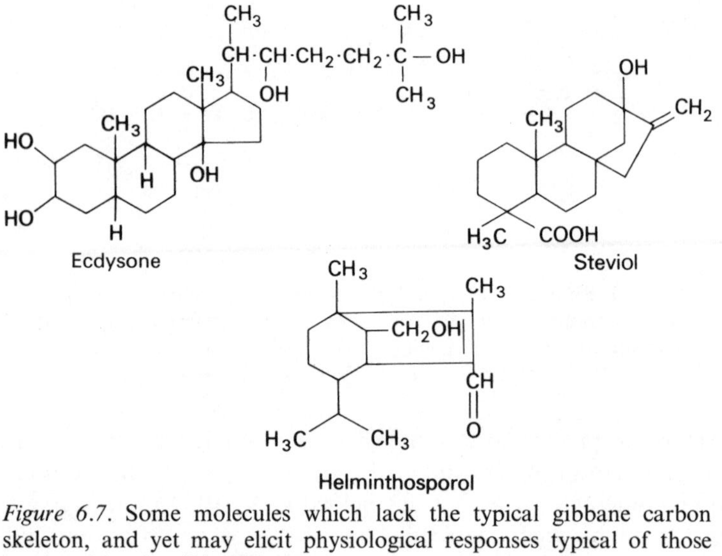

*Figure 6.7.* Some molecules which lack the typical gibbane carbon skeleton, and yet may elicit physiological responses typical of those produced by the gibberellins.

for a response in plants must be based upon differences in their molecular structures. This may, in turn, be related to a necessity for a good 'fit' between gibberellin and species-specific receptor surfaces. On the other hand, there is evidence that gibberellins can be readily interconverted within plants (see also p. 28). This means that the activity or inactivity of any one gibberellin in a particular species may not be determined directly by the shape of the receptor surface, but rather by the ability of enzymes present in that species to convert the applied gibberellin into a form of gibberellin able to link with the receptor molecule. In this same connection, it is probable that many of the gibberellins which have been isolated and identified from plants do not themselves function as hormones, but rather that they represent stages in a biosynthetic sequence leading to the formation of an active gibberellin (see Fig. 1.15).

What evidence is at hand, and it is very slight, suggests that gibberellins, as well as auxins (see above) and cytokinins (below), interact with their site of action by loose physical, and probably noncovalent, bonds.

## Cytokinins

Because natural cytokinins all appear to be based upon 6-amino-purine (adenine), well over a hundred adenine derivatives have been examined in bioassays specifically chosen to reveal cytokinin activity (i.e., to see whether they induce biochemical, physiological and morphological responses similar to those produced by kinetin). In general, it has been found that the side chain attached to the 6-position can be of various types without destroying cytokinin activity, provided that it has a nonpolar nature. A comparison of the side chains of kinetin, $\gamma,\gamma$,dimethylallyl-aminopurine, zeatin and benzyl adenine (Figs. 1.18 and 1.20) indicates the variations possible in the nonpolar side chain. Where the side chain does not contain a ring system, the presence of a double bond usually increases activity, and maximum activity is conferred when the side chain contains five carbon atoms.

Results obtained by Kende *et al.* have suggested that, similar to auxins and gibberellins, cytokinins bind with a cellular receptor site by loose, noncovalent, bonds. This conclusion was reached after the observation that benzyladenine was easily washed out of cells into which it had been incorporated, and in which a physiological response was taking place. The relatively wide range in side-chain size and structure possible in cytokinins based upon adenine, supports the concepts of a rather loose physical association between hormone and receptor site.

## Other hormones

So far as ethylene is concerned, only limited knowledge exists of the essential features of the molecule for activity. It is, of course, a very simple compound in any case. However, certain 3- and 4-carbon chain compounds that have end-group double bonds do exhibit weak ethylene-like activity. It appears that a terminal $=CH_2$ is essential for hormonal activity, and that, given that feature, the smaller the molecule the greater the activity.

Abscisic acid (ABA) possesses an asymmetric carbon atom, and can, consequently, exist as either the $(+)$ or $(-)$ enantiomer (Fig. 1.23). Naturally occurring ABA is always the $(+)$ enantiomer, but despite the fact that some studies with synthetic ABA appeared to show that the $(+)$ optical isomer is more active than the $(-)$ form, recent work by Cornforth and Milborrow indicates that the $(+)$ and $(-)$ enantiomers of ABA are equally effective inhibitors of plant growth. In addition to optical isomerism, ABA also exhibits geometric isomerism around the terminal (no. 2) double bond of the side chain (Fig. 1.23), and here there have been reports that 2-*cis*-ABA has greater growth inhibitory properties than 2-*trans*-ABA, but with both showing activity (Fig. 1.22). However, recent work has demonstrated that 2-*trans*-ABA is not itself biologically active, but is transformed to the active *cis* form in the presence of ultra-violet light. Apart from a possible difference between inhibitory activities of the stereoisomers of ABA, only limited studies have yet been made of ABA analogues to identify features of its molecular structure which are essential for its activity as a plant growth inhibitor. These have suggested that the *cis*,-*trans*-2,4-pentadienoic residue is essential for activity as a plant growth inhibitor, but that the $\alpha,\beta$-unsaturated ketone group in the cyclohexenyl moiety may not be necessary for such activity.

## Mechanical properties of cell walls and hormone-induced cell enlargement

As we saw earlier, a very characteristic effect of auxins and gibberellins is that of inducing cell enlargement. Thus, increased elongation growth in coleoptiles or internodes in the presence of either of these classes of growth hormone, is due to enhanced enlargement of individual cells along the longitudinal axis of these plant organs. Now, each plant cell is contained within a cell wall, and an increase in cell size can occur only if the wall stretches. Even though new cell-wall material is laid down during cell enlargement, the existing wall must expand to accommodate the swelling protoplast.

Plant-cell enlargement is principally a result of vacuolation. That is, although new protoplasm is synthesized during cell enlargement, the uptake of water into the developing vacuole contributes the major component of cell enlargement growth. In a fully vacuolated cell, the living contents are normally distributed as a thin layer just inside the cell wall. In other words, the protoplast becomes inflated by the uptake of water in a manner analogous to the blowing up of a balloon.

### Water relations in enlarging plant cells

Water uptake by plant cells is an osmotic process. This involves the physical diffusion of water molecules across cell membranes which are differentially permeable (i.e., they allow free passage to water but not to other molecules present as solutes in the water). Water tends to diffuse from a region of higher to one of lower *water potential* ($\psi$, the Greek 'psi'), so that the more negative the water potential in a region, the greater is the tendency for water to diffuse to that region. Pure water (i.e., with no solutes) has zero water potential ($\psi = 0$), and an aqueous solution has a negative water potential ($\psi_s < 0$). Consequently, the presence of solutes in the vacuole depresses the water potential of the cell sap ($\psi_v$) and increases the tendency for influx of water from outside.

In addition to the effect of solute concentration in the vacuole, the pressure exerted upon the protoplast by the cell wall also plays a major role in determining the water potential of plant cells. The cell wall pressure is equal in magnitude, but opposite in sign, to the turgor pressure ($\psi_t$) of the protoplast. To summarize, $\psi_s$ is a negative quantity while $\psi_t$ is usually positive. Thus, from the commonly used water-relations equation for plant cells

$$\psi_v = \psi_s + \psi_t,$$

it can be seen that an increase in concentration of solutes in the vacuole (which depresses even further the always negative value of $\psi_s$), at the same time lowers the water potential of the cell ($\psi_v$) and favors water uptake. On the other hand, an increase in $\psi_t$ (positive value) will also increase $\psi_v$ and result in a reduced tendency for water to enter the vacuole.

With a cell that is not growing, and which is contained by a rigid cell wall, there is no net change in water content of the protoplast when $\psi_v = 0$. Clearly, although the protoplast of a turgid cell has a $\psi_v$ of zero, the osmotic concentration of its contents is such that $\psi_s$ is still strongly negative. This means that it is principally the increased positive value of $\psi_t$ which has brought $\psi_v$ up to zero and halted water uptake.

As stated above, $\psi_t$ is equal in value to wall pressure. Thus, the limitation to further water uptake and protoplast swelling in a mature cell is provided by the rigidity, or inextensibility, of the cell wall. Now, during the enlargement of younger cells, further increase in cell size does not appear to be a consequence of a decrease in $\psi_s$ (i.e., by a rise in concentration of the cell contents). In fact, there is considerable experimental evidence that during cell enlargement the cell contents become *less* concentrated (i.e., $\psi_s$ rises), rather than more concentrated. This suggests, therefore, that cell expansion growth occurs as a result of changes in the mechanical properties of the cell wall, which in turn determine wall pressure. The rigidity of the cell wall is reduced, or, put another way, the cell wall becomes more *extensible*. Cell wall extensibility may be of two types: (i) *elastic*, and (ii) *plastic*. Elastic extension of the cell wall would not constitute true growth ('an irreversible increase in size'), for, as the term suggests, elastic extension is reversible. On the other hand, plastic extension is permanent and irreversible. To use the balloon analogy again, on letting the air out of a previously inflated balloon one sees it shrink back to a size *almost* as small as it was originally. Thus, one can distinguish between two components in the stretching of the balloon rubber during inflation; the one being reversible (or elastic), and the other irreversible (plastic).

### Hormonal effects on cell walls

Factors such as auxins or gibberellins which enhance cell enlargement, must, either directly or indirectly, affect the mechanical properties of cell walls. A great deal of work has been done to

**144**

evaluate the influence of auxins on cell walls, and more recently gibberellins have been studied in the same connection. However, despite the fact that both auxins and gibberellins promote cell enlargement (see chapter 2), these two classes of hormone appear to differ in their effects on cell walls. Because of this difference, and the more detailed information available for auxin effects, we will consider first the relationship between auxin and cell-wall properties, and later on indicate how these differ from those of gibberellins.

*Auxin effects.* Very soon after it was discovered that auxin induced cell elongation in coleoptiles (see chapter 1), it was proposed by Heyn (1931) that this was a result of auxin influencing cell-wall plasticity. Heyn's proposal was based upon experiments in which fixed horizontal *Avena* coleoptiles were bent by applied weights. When the weights were removed, then the coleoptiles tended to 'spring back' but never fully recovered their original position. Clearly then, the coleoptiles showed two sorts of mechanically imposed deformation; elastic (reversible) and plastic (irreversible). When treated with IAA, plastic deformation of the coleoptiles was greatly increased (Fig. 6.8). Now, the mechanical properties of a whole coleoptile reflect principally those of individual cell walls, and IAA thus appeared to have increased cell-wall plasticity. The capacity of auxins to cause cell-wall loosening has been amply confirmed in more recent years. In addition, close correlations have been observed between the effect of auxin on growth and on cell-wall plasticity (Fig. 6.9).

Because of the observed plasticizing effect of auxins on cell walls, it was at one time suggested that the primary point of auxin action lies in the cell wall, rather than in the cytoplasm or nucleus. However, although cell enlargement is a common response to auxin, a number of other typical auxin effects do not obviously involve cell-wall stretching (e.g., initiation of cell division in callus cultures, in root primordia initiation and in the cambium). It is probable that changes in cell-wall mechanical properties are initiated by certain metabolic events within the protoplast which can be activated by auxins.

The structure of plant cell walls has been studied in some detail, and we can consequently now begin to understand how auxin could influence its physical properties and thereby cause cell enlargement. The primary walls of young plant cells, which are capable of growing, consist of *cellulose*, *pectic substances*, *hemicelluloses*, *proteins*, and small quantities of *lipids*. Cellulose is a polysaccharide, each

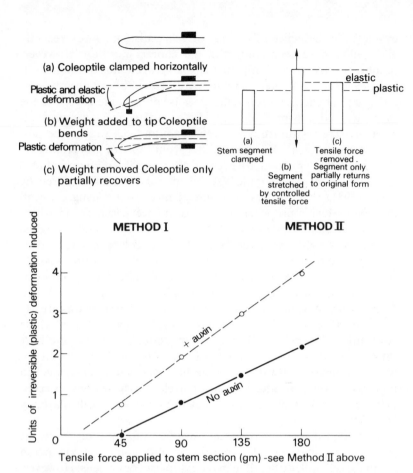

*Figure 6.8. Above.* Two methods for measuring the plasticity of plant cell walls. *Below.* Increased plasticity induced by the addition of an auxin to hypocotyl segments.

(From J. A. Lockhart, C. Bretz, and R. Kenner, *Ann. N.Y. Acad. Sci.* **144** Art I, 19–33, 1967.)

molecule of which consists of a $\beta\,(1 \rightarrow 4)$ linked chain of one thousand or more glucose residues. Both pectic substances and hemicelluloses are also polysaccharides, but in contrast to cellulose these are built up from residues of several different sugars, including arabinose, galactose, glucose, mannose and xylose, and in addition residues of galacturonic acid. Proteins and lipids comprise only a very small proportion of the total cell-wall materials.

Rigidity of the cell wall is conferred mainly by the properties of the noncellulosic components which together form the cell-wall *matrix*. Embedded in the matrix is a mesh of interwoven *microfibrils* which are constituted entirely of cellulose. The matrix is, as has been described, formed of chains of a heterogeneous assemblage of polysaccharides and proteins. It is the strength of chemical bonds, both

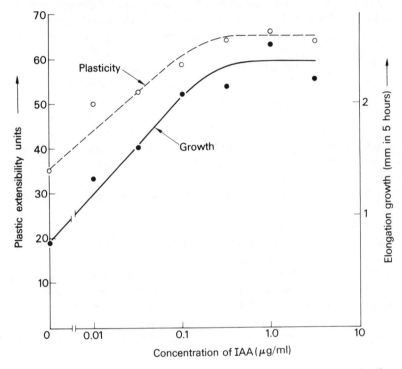

*Figure 6.9.* Parallel effects of auxin (IAA) on elongation growth and cell wall plasticity in oat (*Avena*) coleoptile segments.
(From R. Cleland, *Ann. N.Y. Acad. Sci.* **144** Art I, 3–18, 1967.)

within and between each of the component molecules of the matrix, which determines the overall cohesive properties of the cell wall. Thus, cell enlargement involves breakage of these bonds, and it appears that hydrolytic enzymes are concerned in this. For example, addition of the enzyme $\beta$-1,3-glucanase causes hydrolysis of $\beta$-1,3 glucose bonds of hemicelluloses in *Avena* coleoptile cell walls, and it has been reported that this enzyme also induces both increased cell-

**147**

wall plasticity and elongation growth. This suggests that cell enlargement is achieved through the wall softening effects of enzymes released from the cytoplasm into the cell wall. Indeed, in *Avena* coleoptiles such an enzyme has been detected, and it appears that IAA increases its level of activity (probably by stimulating synthesis of hydrolytic enzymes on the ribosomes of the cytoplasm). However, more recent research has shown that neither $\beta$-1,3 glucanase, nor $\beta$-1,4 glucanase, will induce elongation growth in oat coleoptiles, which tends to reduce the likelihood that such enzymes are concerned in cell enlargement.

The role of neither pectic substances nor proteins in cell walls is fully understood. Until a few years ago, it was thought that auxin effects on pectic components provided the basis for auxin-induced cell-wall softening, but it is now realized that other components of the cell wall are also concerned. It is probable, nevertheless, that pectic substances do contribute to the maintenance of cell-wall structure, principally by them binding with inorganic calcium, but the weight of evidence now indicates that auxin effects on cell walls are mediated by enzymes which hydrolyse the various polysaccharides present. Similarly, although cell walls contain a number of non-enzymic proteins (including an unusual one that contains a high percentage of the amino acid, hydroxyproline) that appear to serve as stiffening agents, there is no evidence to suggest that auxins modify the properties of these in any way which would lead to wall softening.

In summary, cell-wall loosening must occur during cell enlargement growth, and this is promoted by auxin, possibly by the hormone stimulating the synthesis of enzymes which are secreted into the cell wall and effect the breakage of chemical bonds in and between the pectic substances and hemicelluloses of the wall matrix. Further, addition of relatively high auxin concentrations to etiolated pea stems increases cellulase activity in the cell walls. Cellulase catalyses the breakdown of the cellulose molecules of the micro-fibrils, which leads to cell enlargement—indicating that the total mechanical properties of the cell wall are governed by the strength of chemical bonds in microfibrils as well as in the matrix, and that auxin can influence both, perhaps by stimulating production and/or secretion of polysaccharide hydrolysing enzymes. The effect of high auxin concentrations on etiolated pea stems may be due to enhanced ethylene synthesis (p. 42), with the ethylene, in turn, inducing cellulase synthesis.

*Gibberellin effects.* Recent investigations have revealed that although both auxins and gibberellins can promote cell enlargement growth, the two classes of hormone appear to differ in their effects on the mechanical properties of cell walls. Whereas it is well established that auxins induce cell-wall plasticization, the effects of gibberellins on cell walls are less clear cut. In fact, it has recently been found that $GA_3$ can induce cell elongation without causing increased cell-wall softening (Fig. 6.10). In the light of this finding, one must apparently

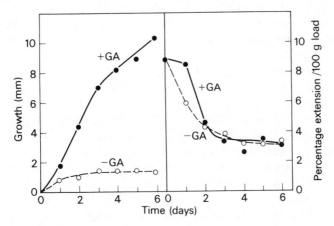

*Figure 6.10.* Stimulatory effect of gibberellic acid on elongation growth, but at the same time with lack of significant effect on cell wall plasticity, in intact cucumber hypocotyls.
(From R. Cleland, M. L. Thompson, D. L. Rayle, and W. K. Purves, *Nature* **219**, 510–511, 1968.)

conclude that gibberellin-induced water uptake and cell enlargement occurs through gibberellin causing an increased concentration of osmotically active solutes in the cells (see cell–water-relations equation on p. 143). This conclusion is not, of course, borne out by observations that the cell contents become more dilute during vacuolation (p. 144). Nevertheless, gibberellins are known to induce the synthesis of enzymes such as $\alpha$-amylase, and proteolytic enzymes (p. 154), and it is possible that the osmotic concentration of cells growing in response to exogenous gibberellin may be increased by the activities of such enzymes (i.e., sugars and amino acids have very much greater osmotic effects than starches or proteins).

# Plant growth hormones and nucleic acid metabolism

Many aspects of plant metabolism are affected, and can be regulated by, the known plant growth hormones. Because those effects are mediated by changes in level or activity of particular enzymes, it is reasonable in the light of modern knowledge of the role of nucleic acids in protein synthesis, to consider the possibility that plant growth hormones operate by influencing nucleic acid metabolism. It is unnecessary in this book to duplicate the excellent descriptions of RNA and protein synthesis given in other modern works (for a summary see Ref. 9 at the end of this chapter), and a general knowledge of these processes will therefore be assumed in the following discussion.

The first demonstration of an effect of a plant hormone on nucleic acids came in 1953, when Silberger and Skoog found that auxin-promoted growth of tobacco-pith cultures was associated with increased levels of both RNA and DNA. Since then, effects of auxins, gibberellins, cytokinins, abscisic acid, and ethylene on RNA metabolism have been observed in a wide range of plant tissues.

In general, promotion of growth by hormones (e.g., by auxins or gibberellins) is associated with an increased rate of RNA synthesis, and growth inhibition (e.g., by ABA) with decreased RNA synthesis. Observed correlations between rates of synthesis of all RNA fractions (messenger, transfer, and ribosomal) and growth, has suggested that proteins (including enzymes) must be synthesized continuously for growth to proceed, and that perhaps growth hormones influence growth through regulatory effects on RNA synthesis.

The pioneering work of Silberger and Skoog demonstrated that auxin-induced enhancement of RNA levels in tobacco-pith cultures occurred prior to an increase in tissue fresh weight, and that as time went by RNA and fresh weight increased in proportion to one another. Subsequently, effects of plant hormones on growth and RNA synthesis have been demonstrated in intact plants, but by far the most fruitful investigations have utilized isolated portions of plants. Greatest attention has been devoted to studies of nucleic acid metabolism in relation to hormonal effects upon (a) elongation growth in coleoptile, hypocotyl, and stem segments, (b) cell enlargement and multiplication in callus cultures, (c) senescence in isolated leaves or leaf disks, and (d) the synthesis of hydrolases in isolated cereal seed aleurone cells.

In this research, advantage has been taken of the availability of a number of compounds which appear to block rather specifically certain steps in the synthesis of nucleic acids and proteins in plant cells. The most useful of these compounds are *actinomycin-D*, which by binding to DNA inhibits the polymerization of RNA in DNA-dependent RNA synthesis, and *cycloheximide* (also known as *actidione*), which does not prevent RNA synthesis but does interfere with protein synthesis on the ribosomes, perhaps due to an inhibition of peptide bond formation and to impairment of the termination or release mechanism in protein formation. Other protein-synthesis inhibitors which have been used are chloramphenicol, puromycin, and 8-azaguanine. Chloramphenicol and puromycin are thought to interfere with protein synthesis at the ribosomal level, whereas 8-azaguanine becomes incorporated into a newly synthesized 'nonsense-RNA' that does not allow normal transcription on the ribosomes. However, it is known that chloramphenicol, puromycin, and 8-azaguanine have various side effects in plant tissues, which makes them less suitable than actinomycin-D or cycloheximide as experimental tools in the study of nucleic acid metabolism. In addition to the blocking of the translation (RNA polymerization on DNA) and transcription (protein synthesis on ribosomes) processes by these inhibitors, experimental manipulation of test systems can be achieved by treating tissues with RNase. This enzyme catalyses the degradation of RNA.

## Growth hormones, nucleic acids, and cell enlargement

Continued synthesis of RNA and protein is essential for continued cell elongation growth. Thus, treatment of an elongating plant organ with increasing concentrations of actinomycin-D causes parallel reductions in RNA content and growth (Fig. 6.11). Moreover, when elongation growth is enhanced by the addition of a growth hormone, the addition of an agent such as actinomycin-D, cycloheximide, or RNase, can result in parallel inhibition of hormone-induced growth and DNA-dependent RNA synthesis (e.g., for auxin and actinomycin-D on soybean hypocotyl segments, Fig. 6.12). Thus, auxin-enhancement of plant cell enlargement appears to occur only when RNA and proteins are being synthesized. Similar relationships between RNA and protein synthesis, and gibberellin-enhanced elongation growth have been inferred from experiments on lettuce hypocotyl elongation and the effects of actinomycin-D and puromycin.

Correlations between rates of RNA and protein synthesis on the one hand, and auxin or gibberellin enhancement of cell enlargement on the other, do not in themselves demonstrate that the mechanism of action of these hormones involves a *direct* effect at transcription of DNA-coded information. It is possible that observed effects of

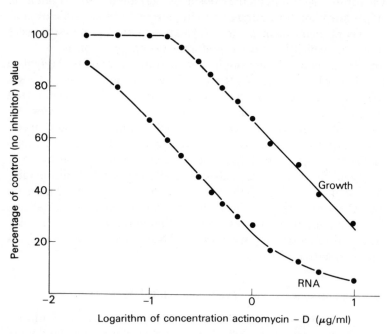

*Figure 6.11*. Parallel inhibition by Actinomycin-D of endogenous elongation growth and RNA synthesis in soybean hypocotyl segments.
(From J. L. Key, N. M. Barnett, and C. Y. Lin, *Ann. N.Y. Acad. Sci.* **144** Art I, 49–62, 1967.)

growth hormones on total RNA content are of no greater significance than their effects on other metabolic processes, so far as identifying the primary point of hormone action.

Regulation of protein synthesis is thought to be achieved principally through differential rates of synthesis of m-RNA (messenger-RNA), in which qualitative, as well as quantitative, differences may occur, depending on which segments of the DNA code are being 'read'. Sequential polymerization of m-RNA of different base-sequences would lead to the appearance of a series of distinctive enzymes. We may presume that developmental changes take place as a result of an orderly sequence of enzyme production, so that great interest lies in establishing whether or not growth hormones regulate m-RNA synthesis. However, the majority of studies of cell enlargement have, to the present time, recorded changes in *total* RNA content occurring in association with hormone-induced growth; only a tiny portion of this would be relatively 'short-lived' m-RNA.

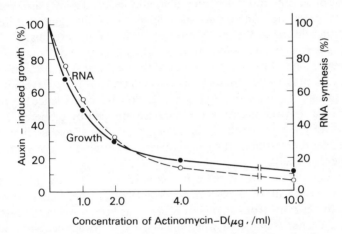

*Figure 6.12.* Parallel inhibition by Actinomycin-D of auxin-induced growth and DNA-dependent RNA synthesis in soybean hypocotyls. (From J. L. Key, N. M. Barnett, and C. Y. Lin, *Ann. N.Y. Acad. Sci.* **144** Art I, 49–62, 1967.)

The technical problems associated with attempts to separate freshly-synthesized new types of m-RNA from the total RNA content of higher plant tissues has prompted research on simpler, but still hormone-responsive, systems than elongating plant organs.

Particularly valuable results have been obtained using isolated barley aleurone tissue, in which hydrolases are synthesized when a gibberellin is supplied. Other research groups have concentrated their attention upon ethylene-induced, and cytokinin or gibberellin-delayed, senescence and abscission processes. More recently, work has commenced on the responses of isolated plant-cell nuclei, or isolated *chromatin* (principally DNA and nucleoprotein), to applied hormones, and it may prove easier to detect any hormone-induced changes in RNA synthesis at the surface of relatively pure cell-free DNA, than it is in the much more complex cellular environment.

## Gibberellin-induced hydrolase synthesis and nucleic acid metabolism

In an earlier chapter (see p. 116, and Figs. 5.6 and 5.7), a general account was given of the effect of exogenous $GA_3$, or embryo-derived gibberellin, on the synthesis of $\alpha$-amylase and other enzymes in cereal seed aleurone cells. Work by various research groups, but particularly those of Varner and Paleg, on gibberellin-induced $\alpha$-amylase synthesis in barley aleurone tissue has yielded especially significant information relating to the mechanism of action of gibberellins.

Isolated aleurone tissue from unsoaked barley seed contains only traces of $\alpha$-amylase, but treatment of the aleurone with a solution of $GA_3$ results in a very marked increase in the $\alpha$-amylase level after a six- to eight-hour lag phase (Fig. 6.13). Conclusive evidence has been obtained that the increase in $\alpha$-amylase induced by gibberellin is a result of *de novo* synthesis of the enzyme. Thus, gibberellins can cause de-repression of the gene concerned with $\alpha$-amylase synthesis in aleurone cells.

Evidence that the increase in $\alpha$-amylase induced by gibberellin occurs as a result of *de novo* enzyme synthesis, rather than of release of active enzyme from an inactive form, has been obtained in several ways. Varner found that when aleurone tissue was incubated with $^{14}C$-phenylalanine in the presence of $GA_3$, the $\alpha$-amylase appearing was labeled with $^{14}C$, suggesting that at least part of the increase in $\alpha$-amylase activity was attributable to newly synthesized enzyme. Similarly, 'fingerprinting' of tryptic digests of purified $\alpha$-amylase, produced in response to $GA_3$ treatment, showed that the entire amylase molecule had been synthesized during the experiment. Proof that *all* the $\alpha$-amylase produced in response to $GA_3$ is synthesized *de novo* from amino acids came from *density-labeling* experiments by Filner and Varner in 1967. These workers incubated barley aleurone cells with $GA_3$, together with either normal water ($H_2O^{16}$) or water containing the heavy isotope of oxygen ($H_2O^{18}$). In aleurone cells, amino acids become available for protein synthesis by hydrolysis of storage proteins. Hydrolysis in the presence of $H_2O^{18}$ results in the liberation of amino acids density labeled with $O^{18}$, and any proteins synthesized from them will also be density labeled (i.e., will be heavier than proteins formed from amino acids released in the

154

presence of $H_2O^{16}$). Filner and Varner discovered with this technique that α-amylase appearing in $GA_3$-treated aleurone cells incubated with $H_2O^{18}$ was of about the theoretically expected one per cent greater density than α-amylase from $H_2O^{16}$-incubated aleurone (Fig. 6.14), clearly indicating that all of the induced α-amylase was newly synthesized from amino acids during incubation with $H_2O^{18}$.

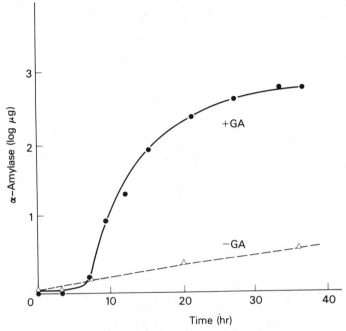

*Figure 6.13.* Effect of gibberellic acid (GA) on the synthesis of α-amylase in barley-seed aleurone tissue. Aleurone plus endosperm pieces were incubated in buffer with (+GA) or without (−GA) gibberellic acid at a concentration of $10^{-6}$ M.
(From Robert E. Cleland, in *The Physiology of Plant Growth and Development*, ed. M. B. Wilkins, pp. 49–81, McGraw-Hill, London, 1969. After J. G. Varner *et al.*, *J. Cell. Comp. Physiol.* **66**, Suppl. 55–68, 1965.)

Synthesis of α-amylase in barley aleurone cells under the influence of gibberellin requires RNA synthesis. Actinomycin-D added during the first six to eight hours of incubation with $GA_3$ (the lag phase which occurs before α-amylase formation begins) completely prevents α-amylase synthesis, but after the lag phase the mechanism is

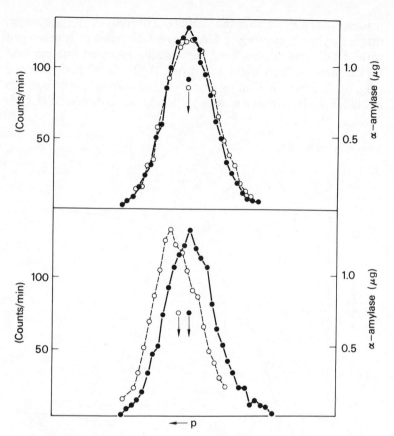

*Figure 6.14.* Evidence by 'density-labeling' that the entire α-amylase molecule synthesized in response to gibberellin treatment of barley aleurone cells is formed from amino acids ($\rho$ = density). *Above.* Coincidence of densities of α-amylase produced in presence of $H_2O^{16}$ (open circles) and highly purified tritiated marker α-amylase (closed circles) prepared by incubating aleurone tissue in $H_2O^{16}$ and lysine-$H^3$. *Below.* Increased density of α-amylase produced in presence of $H_2O^{18}$ (open circles) compared with the purified marker α-amylase of normal density (closed circles).

(From P. Filner, and J. Varner, *Proc. Nat. Acad. Sci.* **58**, 1520–1526, 1967.)

completely insensitive to actinomycin-D. This suggests that all the m-RNA required for α-amylase formation is synthesized during the first six to eight hours, and that thereafter α-amylase synthesis can proceed using the m-RNA already formed as template. As would be expected, addition of the protein-synthesis inhibitor cycloheximide at any time during the actual formation of α-amylase stops further synthesis of the enzyme. However, withdrawal of gibberellin even after the lag-phase period of m-RNA synthesis results in cessation of α-amylase synthesis, and this suggests that gibberellin is required for some reason in addition to that of inducing de-repression of the gene concerned with α-amylase, but what that reason is remains obscure at present. It is possible, on the other hand, that uninterrupted synthesis of a particular m-RNA is essential for continued gibberellin-regulated α-amylase synthesis, but that present-day techniques are inadequate to detect the functional RNA. This suggestion assumes, perhaps incorrectly, that the failure of actinomycin-D to inhibit α-amylase synthesis after eight hours' incubation with $GA_3$ is due to the inability of the RNA-synthesis inhibitor to prevent polymerization of specifically that RNA fraction required for continued α-amylase synthesis.

The growth-inhibitory hormone, ABA, also inhibits $GA_3$-induced α-amylase synthesis, while at the same time only slightly reducing RNA synthesis, and not affecting at all the total incorporation of amino acids into protein. The inhibitory effect of ABA on α-amylase formation appears to be fairly specific for this enzyme, and it is only partially overcome by increasing the concentration of $GA_3$ present. Chrispeels and Varner (1967) have suggested that ABA inhibits the synthesis of a specific RNA fraction required for α-amylase synthesis.

It is clear, therefore, that gibberellins can cause the de-repression of the gene for α-amylase in barley aleurone cells, and that ABA prevents the expression of the same gene. It is not yet possible, however, to decide whether gibberellins or ABA operate directly at the transcriptional level (i.e., at the gene) or at the translational level (during protein formation). Until information is obtained concerning any effects of gibberellins and ABA on specific RNAs, or evidence for specific regulation at the translational level, then open minds must be kept on the problem of gibberellin and ABA action mechanisms in the barley aleurone system.

### Hormonal effects on nucleic acid metabolism in relation to senescence and abscission

The processes of senescence and abscission (see chapter 5), particularly in leaves, have received study in relation to the effects of growth hormones, and certain of the results obtained have relevance to the problem of the relationship between nucleic acid metabolism and hormone action.

A principal characteristic of a senescing organ such as an old, or detached, leaf, is a gradual fall in protein and RNA content (Figs. 5.8 and 5.9). The proteins of a leaf are normally in a state of dynamic equilibrium with the pool of available amino acids. In other words, a continuous 'turnover' in protein is always going on, and the protein content of a healthy green leaf represents a 'steady state'. Thus, any fall in protein level, such as occurs at senescence, can be due to one or more of three possibilities: (a) an increased rate of protein breakdown, (b) a decreased rate of protein synthesis, (c) translocation of amino acids out of the leaf to other parts of the plant, with no change in the rate of protein breakdown but a consequent reduced rate of protein synthesis. The third possible cause of protein loss (increased amino-acid export) can be discounted, for senescence occurs in detached leaves, or in leaf disks, under conditions where amino-acid export is impossible. In fact, amino acids have been observed to *increase* in concentration during senescence of leaf disks and detached leaves. Similarly, protease activity has been found to remain constant during the senescence of leaf tissues, so that it appears that the fall in protein in senescing leaves is attributable to a failure in the protein-synthesis mechanism. Delay of yellowing in detached leaves by gibberellins, cytokinins, or auxins, and its acceleration by ABA or ethylene (see chapter 5), has been found to be paralleled by effects of these growth hormones on rates of protein synthesis.

It therefore appears, once again, that hormonal effects in senescence are mediated through effects on RNA synthesis. Analysis of nucleic acids in relation to senescence, and the effects of growth hormones, have revealed that those hormones which delay senescence allow maintenance of RNA synthesis (e.g., see Fig. 5.9), whereas ABA or ethylene depress the rate of RNA synthesis. Concurrent treatment of a detached leaf with a senescence-delaying growth hormone and an RNA- or protein-synthesis inhibitor, such as actinomycin-D or cycloheximide, results in blocking of the effect of the hormone. Nevertheless, available evidence does not allow any conclusions to be drawn as to the mechanism whereby plant growth hormones affect nucleic acid and protein metabolism during leaf senescence.

Abscission of leaves and fruits involves the dissolution of cell walls in the separation layer (Fig. 5.12). This occurs as a result of a rise in cellulase activity in the separation layer cells. This enzyme catalyses the hydrolysis of the cellulose microfibrils of the cell walls (p. 145). The production of cellulase is preceded by a rise in RNA synthesis in the abscission zone (but not in other regions of the petiole). Ethylene is known to be the hormone which induces abscission in senescent leaves (p. 122), and exposure of old leaves to this gas results in an immediate enhancement of RNA synthesis in the abscission zone, followed by a rise in the rate of protein, including cellulase, synthesis. Consequently, actinomycin-D or cycloheximide prevent ethylene-induced abscission. Treatment of petioles with actinomycin-D several hours after their exposure to ethylene does not prevent the ethylene-induced effects on RNA and protein synthesis, but cycloheximide inhibition occurs for somewhat longer.

It seems clear that ethylene regulates the rate of RNA synthesis in connection with cellulase production in cells of the abscission zone but, once again, we cannot reach definitive conclusions as to whether ethylene acts *directly* on nucleic acid and protein-synthesis mechanisms.

### Growth hormone effects on isolated nuclei and chromatin

The work discussed above concerning nucleic acid metabolism in plant organs or tissues has, so far, failed to provide the information needed to identify the point of primary action of growth hormones in the cellular machinery. It is unfortunate that this is the case, for much of the research involved has been of intellectual and technical excellence.

Should plant growth hormones have direct effects within the nucleus on RNA-polymerization processes, then these should be somewhat simpler to detect in experiments using isolated, cell-free nuclei or chromatin. Such experiments involve the incubation of isolated nuclei or chromatin, with necessary RNA-precursors (ATP, GTP, CTP and UTP), which can be isotopically labeled to facilitate analysis of newly formed RNA. Conflicting results have been obtained in such experiments, for some have appeared to demonstrate auxin-enhanced RNA synthesis in isolated plant nuclei, whereas in other cases no effect of auxin was found. The cause of variable results being obtained with isolated nuclei is not yet clear, but at least two possible explanations present themselves. These are that, (a) in some experiments at least, bacterial contamination of the isolated nuclei contributed significantly to the total quantity of RNA synthesized, and (b) different procedures for isolating the nuclei had marked effects on the response of these to added auxin. One aspect of the isolation method adopted which seems particularly important is whether or not auxin is added prior to the nuclei being separated from cytoplasmic constituents of cells, for only where the cells were exposed to auxin before the nuclei were removed did auxin-enhancement of RNA synthesis occur in the isolated nuclei.

Similar results to those obtained for auxin-effects on RNA synthesis in isolated nuclei have occurred in experiments with gibberellins. For example, Johri and Varner in 1968 found that nuclei isolated from cells of dwarf pea in the presence of $10^{-8}$ M $GA_3$, subsequently incorporated up to 80 per cent more ${}^3$H-labeled nucleotide into RNA when incubated in the presence of $10^{-8}$ M $GA_3$, than they did if isolated and incubated in the absence of $GA_3$. Thus, as with auxin, the gibberellin must be supplied before final separation of the nucleus from other cellular components if any effect is to be shown on RNA polymerization. Fractionation of the RNA formed in the nuclei revealed that $GA_3$ affected the kind, as well as the quantity, of RNA synthesized. The whole intact cell did not, however, need to be present for $GA_3$-enhanced RNA synthesis, but the later $GA_3$ was added during the process of isolation of nuclei, then the smaller was the response.

It may, therefore, be that auxin or gibberellin must react with some cytoplasmic factor or factors, following which an influence is exerted on RNA synthesis in the nucleus. In other words, that the hormone does *not* act directly at the transcriptional level. On the other hand, the failure of nuclei isolated in the absence of added hormone to respond to hormone supplied subsequently, may be a result of loss of some essential hormone-sensitive factor from nuclei during their isolation. This second possibility is supported by some recent research of Matthysse (1968) with chromatin material, the results of which are discussed in the next paragraph.

In general, addition of auxin or gibberellin to isolated chromatin does not result in a higher level of RNA synthesis, unless, as with whole nuclei, the hormone is added before, or at least during, the actual separation of chromatin from the cell. Similarly, RNA synthesis on isolated chromatin is doubled if the plants are treated with ethylene before chromatin extraction, and the types of RNAs produced are also different from those formed on chromatin from untreated seedlings. In the other direction, ABA-treatment of plants, or the addition of ABA during chromatin isolation, reduces the synthesis of RNA *in vitro* on the isolated chromatin. It is clear, therefore, that chromatin obtained from hormone-treated cells is different from that prepared from untreated cells. Work by Cherry, Key, O'Brien and associates at Purdue University has indicated that auxin pre-treatment either increases the RNA polymerase of isolated chromatin, or increases the template available for transcription. It is not yet possible to decide between these possibilities, but results obtained by Matthysse (1968) may have significance here. Matthysse found that nuclei isolated from tobacco and soybean in the absence of auxin failed to respond to auxin *in vitro*. However, the addition of a factor obtained from pea, tobacco or soybean nuclei to the isolated nuclei allowed the latter to respond to auxin *in vitro* by synthesizing more RNA. The nature of the extracted factor is not known, but it appeared to be protein. Addition of only the factor, or of auxin alone, to pea chromatin in the presence of RNA polymerase from *E. coli* had no effect on RNA synthesis, but the supply of *both* the factor and auxin to the chromatin system doubled the rate of RNA synthesis. It is possible that the presence of auxin before and during the separation of chromatin from the cell prevents the loss of this hormone-sensitive factor from the nuclear material.

## Translational control by growth hormones, and incorporation of cytokinins into transfer-RNA

In the introductory comments to this chapter (p. 134), mention was made of efforts to locate within cells radioactive exogenous plant growth hormones. Autoradiography of tissues treated with auxins or cytokinins has yielded some information (Fig. 6.1), but by far the most interesting results have been obtained in studies of extracts of tissues treated with labeled cytokinins.

Cytokinins, like auxins, ethylene, and gibberellins, stimulate nucleic acid and protein synthesis. The purine nature of natural cytokinins suggests that members of this class of hormone become, in some manner, incorporated into plant nucleic acids. If this does occur, then it may be envisaged that their mechanism of action is related to their inclusion in nucleic acids. Studies of the fate of exogenous cytokinins in plant tissues have revealed that they are degraded in plant cells to yield a number of low molecular weight substances. The purine ring is, however, subsequently utilized for a variety of synthetic reactions, including RNA formation.

The occurrence of cytokinins in RNA was first reported in 1964 by J. E. Fox of the University of Kansas. It has been found that incorporated cytokinins are almost exclusively restricted to certain fractions of soluble- (principally transfer-) RNA, and these discoveries make tempting the proposition that cytokinins operate through their incorporation into certain types of t-RNA. The isolation of IPA and zeatin ribosides (Fig. 1.21) from t-RNA of yeast and *Zea mays* lends support to this concept.

It has been found that cytokinins can be incorporated into at least two types of specific t-RNA: serine t-RNA and tyrosine t-RNA, but not into others (e.g., arginine, glycine, phenylalanine, valine, and alanine t-RNA do not contain cytokinins). The full significance of such relative specificity is still not clear to us. However, the stimulating discovery in 1966 by Zachau and co-workers that IPA is localized in yeast serine-transfer RNA at a position adjacent to the anticodon loop, perhaps provides an important clue. Fuller and Hodgson (1967) have reasoned that a 6-substituted purine (all natural cytokinins are of this chemical nature) positioned adjacent to the anticodon loop would allow maintenance of the correct spatial arrangement of the t-RNA, and so prevent an incorrect triplet of nucleotides being recognized by the codon of m-RNA on the ribosome. In fact, it seems likely that the presence of a molecule of the cytokinin type is essential for normal codon–anticodon interaction between t-RNA and m-RNA on the ribosome. Evidence for this has come from observations that certain inactive cytokinin analogues, when incorporated into serine t-RNA, had no effect on the acceptance and transfer of 'activated' serine but did interfere with binding of the charged serine-t-RNA to the m-RNA-ribosome complex.

The concept of translational control by cytokinins is very attractive, but unfortunately one cannot feel at all sure that the hormonal effects of cytokinins are brought about by their incorporation into t-RNA. In the normal biosynthetic sequence, modification of bases in t-RNA probably takes place after the primary structure of the polynucleotide has been laid down. Thus, the characteristic side chain on carbon-6 of the adenine moiety of a cytokinin (p. 141) is presumably attached *after* the adenine portion is incorporated into t-RNA. This makes it difficult to see how substances such as kinetin, BAP, IPA and zeatin can be incorporated as such into newly formed t-RNA. Adenine is the critical and specific portion of molecules active as cytokinins, and it is possible that exogenous cytokinins become degraded in the cell, following which released adenine is scavenged and utilized in t-RNA synthesis. This possibility is supported by the fact that although the *cis*-isomer of ribosyl zeatin occurs in *Zea mays* seed t-RNA, ethanol extracts of the same seeds contain the *trans*-isomer of zeatin, indicating that zeatin is not a precursor of t-RNA. Further, Kende and Travares (1968) found that $^{14}$C-labeled 6-benzylamino-9-methyl-purine was active as a cytokinin, yet due to the masking of the 9-position of the purine residue with a methyl group it was not incorporated into RNA at all. These results argue very strongly against the idea that the mechanism of cytokinin action depends upon incorporation into t-RNA, and no essential experimental demonstration has been made to permit us to believe that cytokinins exert their physiological effects through regulation of protein synthesis at the translational level. Similarly, there is no good evidence that any other of the classes of known plant growth hormones operate through direct incorporation into nucleic acids.

Nevertheless, some recent work by Chen and Osborne (*Nature*, **226**, 1157, 1970) on the effects of $GA_3$ and ABA on DNA, RNA, and protein synthesis in wheat embryos suggests that in this system the growth hormones exert translational control. Only when embryo DNA was in a derepressed condition (not until after 12 hours of germination) and RNA synthesis could occur, did $GA_3$ and ABA have their respective stimulatory and inhibitory effects on RNA synthesis. Neither $GA_3$ nor ABA affected the time of derepression of DNA. This result argues against a role for these hormones at gene transcription. However, $GA_3$ enhanced, and ABA suppressed, the rate of protein synthesis on the ribosomes well before the end of the 12 hour period during which no RNA synthesis took place. This indicates that the hormones affected protein synthesis by regulating the utilization of m-RNA already present in the embryo (presumably stored since the seed was shed from the parent plant), rather than by initiating the synthesis of new m-RNA.

## Some conclusions

Much hard work and not a little inspiration over a period of some 90 years, has led us to the point where we know (a) that plant growth hormones exist, (b) that there are at least three classes of *growth promoters* (auxins, cytokinins, gibberellins), and two of *growth inhibitors* (ABA, ethylene), and (c) that although each class of growth hormone has its own chemical characteristics and physiological roles, nevertheless the effects of the different classes of hormone are not distinct from one another, and interactions between different hormones are vitally important in the overall control and integration of growth and differentiation in the plant.

The reading of this book should have 'put you in the picture' so far as current knowledge and concepts allow. Undoubtedly, it is likely that there are aspects of hormonal control in plants which have completely escaped our notice up to the present time, and it is also possible that there are other, completely unknown, types of hormonal substances at work—we are at least aware of the transmissible floral stimulus, even if its nature remains obscure. Even to attempt to foresee future developments in this field of research is daunting, so rapid is the pace and violent the flux of opinions.

Finally, there is the basic problem of the mechanism of action of plant growth hormones. The experimental evidence discussed in the last few pages of this chapter leads to the conclusion that plant hormones can exert regulatory effects on nucleic acid metabolism, which means that the quantities and types of proteins and enzymes which are formed may alter. Nevertheless, as pointed out several times earlier, no direct evidence exists for the direct participation of a plant growth hormone at transcription. This may be attributable to nonavailability of such evidence (i.e., that these hormones do not directly control RNA synthesis), or to purely technical deficiencies of the experimental procedures followed. As an alternative to transcriptional control, it may be considered that plant hormones serve to regulate at the translational level (i.e., at protein synthesis). The occurrence of cytokinins in a region of certain t-RNAs adjacent to the anticodon loop is strongly suggestive of translational control, but for the reasons discussed in the preceding section of this chapter, even this may not necessarily be taken to mean that cytokinins directly regulate protein synthesis, or that their hormonal effects are dependent on incorporation into t-RNA.

Evidence interpretable in terms of translational control by growth hormones, obtained in studies of the effects of $GA_3$ and ABA on nucleic acid and protein synthesis in germinating wheat embryos, was discussed on page 163, and we must therefore keep open the possibility that at least certain categories of plant growth hormones, and under certain conditions, do operate by a direct regulation of protein synthesis at ribosomal level.

Central to the question of the mechanism of plant growth hormone action is the ever present problem of separating cause from effect in biological responses. Thus, are the increases in RNA and protein synthesis, or cell wall loosening, which may follow hormone treatment, causal events in hormone-regulated processes? Or are they consequences of an earlier initial reaction between hormone and another, or other, cellular components? Very recent microscopic studies of the early time-course of IAA-enhanced cell elongation in *Avena* and *Zea mays* coleoptiles have yielded results which argue against the idea that the primary effect of auxin is to stimulate DNA-dependent RNA synthesis. Thus, the cell enlargement response to auxin commences either after a latent period of about ten minutes (which could be long enough for enhanced RNA and protein synthesis to take place), or, under appropriate conditions, *immediately* the auxin is applied. That is, auxin can enhance cell enlargement in circumstances which do not allow time for RNA and protein synthesis to occur. Other examples of fast responses to auxin are known, which also would appear to be too rapid to be explained in terms of a primary effect of the hormone on the processes of RNA and protein synthesis (e.g., promotion of protoplasmic streaming within twenty seconds, and effects on plasma membrane permeability to ions and non-electrolytes). It has consequently been suggested (reference No. 6 at the end of this chapter) that the primary interaction of auxin is with the plasma membrane.

The most honest, and the safest, conclusion to reach on the subject of the mechanism of plant hormone action, is the negative one. We just do not know. All the studies made so far of nucleic acid metabolism in relation to plant-hormone action, have largely assumed that the model of nucleic acid–protein synthesis erected from results of research with bacteria by molecular biologists is applicable to higher plant cells as well. If this should prove not to be the case, and very recent research indicates certain differences in nucleic acid metabolism between higher organisms and bacteria, then undoubtedly research into the mechanism of plant growth hormone action will go on even more vigorously than at present, taking into account new knowledge of the system which is being affected by the hormone. There is unquestionably plenty more to do, and time enough for you, the student, to become one of the researchers doing it!

## Further reading list

1. Fox, J. E. (ed.) *Molecular Control of Plant Growth*, Prentice-Hall International, 1968.
2. Fox, J. E. 'The Cytokinins', in *The Physiology of Plant Growth and Development* (ed. M. B. Wilkins), pp. 85–123, McGraw-Hill, London, 1969.
3. Frederick, J. F. (consulting ed.) 'Plant Growth Regulators', *Annals New York Academy of Science*, Vol. 44, art. 1, pp. 1–382, 1967.
4. Key, Joe L. 'Hormones and Nucleic Acid Metabolism', *Ann. Rev. Plant Physiol.* **20**, 449–474, 1969.
5. Overbeek, Van. J. 'The Control of Plant Growth', *Scientific American* **219**, (1), 75–81, 1968.
6. Rayle, D. L., Evans, M. L. and R. Hertel. 'Action of Auxin on Cell Elongation', *Proc. National Academy of Sciences* **65**, 184–191, 1970.
7. Skoog, F., and D. J. Armstrong. 'Cytokinins' *Ann. Rev. Plant Physiol.* **21**, 359–384, 1970.
8. Thimann, K. V. 'Plant Growth Substances; Past, Present, and Future', *Ann. Rev. Plant Physiol.* **14**, 1–18, 1963.
9. Wareing, P. F. and I. D. J. Phillips. Chapter 13 in *The Control of Growth and Differentiation in Plants*, Pergamon Press, Oxford, 1970.

# Index

Root initiation, 23, 70–72
*Rumex* (dock), 119

*Samolus parviflorus*, 48–49, 95
*Scabiosa*, 65
Secondary thickening, 55–58
Senescence:
    chlorophyll levels during, 119–122,
      158–159
    of fruits, 105, 122–123
    of leaves, 118–122, 157–159
    nucleic acid metabolism during,
      157–159
    protein levels during, 119–122,
      158–159
    of whole-plant, 123–124
Separation layer, 124–126, 132
Sesquiterpene, 28, 37
Shoot growth, 45–67
Short-day plants, 92
*Silene pendula*, 98
*Solanum tuberosum* (potato), 108,
    112
Soybean, 98, 152–153, 161
Spinach, 124
Statocyte, 86
Statoliths, 86
Sterile culture of plant cells and
    tissues, 30–32
Sterols, 26
Steviol, 140
Stratification of seeds, 112
Structure—activity relationships of
    growth hormones, 133, 135–142
    of abscisic acid, 142
    of auxins, 136–139
    of cytokinins, 141–142
    of ethylene, 142
    of gibberellins, 140–141
Sub-apical meristem, 46–49

*Taraxacum* (dandelion), 119
Tendrils, 74
Terpenoids, 26–28
Thigmonastisism, 74
Thigmotropism, 74
Thiocarbamates as auxins, 137, 139
*Thuja placata* (Western red cedar), 71

Translocation:
    of abscisic acid, 38–39
    of auxins, 17–23
    of cytokinins, 35, 66
    of ethylene, 41
    of gibberellins, 28–29
    of nutrients, 21, 62–65
2,3,6-Trichlorobenzoic acid, 137–139
2,4,5-Trichlorophenoxyacetic acid
    (2,4,5-T), 136–137
*Trifolium* (clover), 74
*Triticum* (wheat), 11, 36, 109, 116–
    117, 163
*Tropaeolum* (nasturtium), 119–120
Tropisms, 73–89
Tryptamine, in IAA biosynthesis, 13
Tryptophan, as precursor for IAA
    biosynthesis, 12–14
Two-point attachment theory of
    auxin action, 138

Vacuolation of cells, 48–49
Vascular cambium, 55, 67, 70
*Vicia faba* (broad bean), 61, 65
Violaxanthin:
    and IAA photo-oxidation, 14
    as precursor of abscisic acid, 38–39
Vitamins, 3–4
*Vitis vinifera* (grape), 102

Water-potential, 143–144
Water relations of plant cells, 143–
    144
*Wolffia microscopica*, 97

*Xanthium pensylvanicum* (cocklebur),
    92–93, 97–98
Xylem, 55–58
Xylose, 146

*Zea mays* (corn, or maize), 22, 33, 80,
    86, 88, 140, 162–163, 165
Zeatin (6-(4-hydroxy-3-methylbut-2-
    enyl)aminopurine), 33–34, 97,
    141, 162–163
Zinc, deficiency effect on auxin syn-
    thesis, 12

## DATE DUE

| | | | |
|---|---|---|---|
| | | | |
| | | | |
| | | | |
| | | | |
| | | | |
| | | | |
| | | | |
| | | | |
| | | | |
| | | | |
| | | | |
| | | | |
| | | | |
| | | | |
| | | | |
| | | | |
| | | | |
| GAYLORD | | | PRINTED IN U.S A. |